Australian Curriculum

ScienceWorld 10

Australian Curriculum edition

Lesley Englert · Peter Stannard · Ken Williamson
BA, DipT BSc, DipEd BSc (Hons), DipEd

This edition published in 2021 by

Matilda Education Australia, an imprint
of Meanwhile Education Pty Ltd
Melbourne, Australia
T: 1300 277 235
E: customersupport@matildaed.com.au
www.matildaeducation.com.au

First edition published in 2012 by Macmillan Science and Education Australia Pty Ltd

Publication data

Authors: Lesley Englert, Peter Stannard and Ken Williamson
Title: ScienceWorld 10 Workbook: Australian Curriculum edition
ISBN: 978 1 4202 2997 4

Publisher: Peter Saffin
Project editor: Inge Note
Illustrator: Chris Dent; Andy Craig and Nives Porcellato
Cover designer: Dim Frangoulis
Text designer: Firefly Education
Production control: Loran McDougall
Typeset in Helvetica, Trade Gothic and Linotype Kaliber by Firefly Education
Cover images: Escalante National Monument, Utah, USA, courtesy of Getty Images/Dian Cook and Len Jenshel; stars background ©Shutterstock/Wolfgang Kloehr

Printed in China by Central
Sep - 2024

Warning: It is recommended that Aboriginal and Torres Strait Islander peoples exercise caution when viewing this publication as it may contain images of deceased persons.

About this book

Are you ever faced with a problem and you don't know how to go about solving it? For example, look at the problem below.

Situation: Every time Sam goes into the garden in spring, she starts sneezing uncontrollably. This does not happen in any other season, and never occurs inside the house. What is in the garden that makes Sam sneeze?

There is no correct or incorrect way to solve problems. This workbook gives you a number of problem-solving strategies to draw upon. The most commonly used strategies are shown below—

1 Break down the problem into manageable parts and then work through them one at a time.
2 Devise a table, chart or graph.
3 Work backwards to eliminate other possibilities.
4 Infer and test.
5 List all the possibilities and select the most likely.
6 Use a rule or formula.
7 Use evidence to reach a logical conclusion.
8 Draw a diagram or make a model.
9 View the problem from different perspectives.
10 Select relevant information.

The longer Tom contemplated the wood pile, the better an electric stove looked.

Sometimes you may use only one strategy to solve a particular problem. In other cases you may have to use more than one strategy.

In the first 10 chapters of this workbook you will be shown how to use the problem-solving strategies listed above. Each strategy is modelled for you on the first page of the chapter and a new problem for you to solve using the same strategy is on the last page of the chapter.

In Chapter 11 you use one or more of the 10 strategies to solve a problem.

Getting to know your workbook

Each chapter begins with the same chapter number and name as found in your textbook.

Sometimes you will see a section called What do you know already? This will show you how much you already know about the work in the chapter. *This is not a test.*

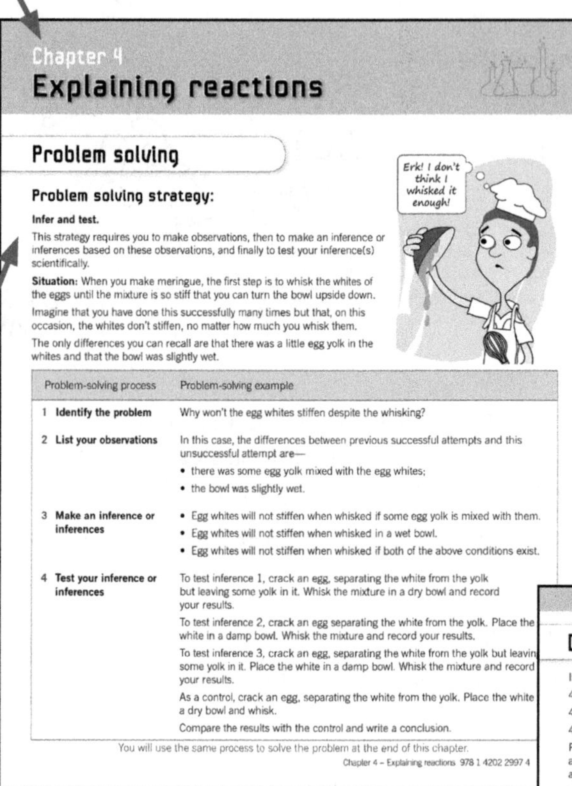

Chapter 4

Explaining reactions

Problem solving

Problem solving strategy:

Infer and test.

This strategy requires you to make observations, then to make an inference or inferences based on these observations, and finally to test your inference(s) scientifically.

Situation: When you make meringue, the first step is to whisk the whites of the eggs until the mixture is so stiff that you can turn the bowl upside down.

Imagine that you have done this successfully many times but that, on this occasion, the whites don't stiffen, no matter how much you whisk them.

The only differences you can recall are that there was a little egg yolk in the whites and that the bowl was slightly wet.

Problem-solving process	Problem-solving example
1 **Identify the problem**	Why won't the egg whites stiffen despite the whisking?
2 **List your observations**	In this case, the differences between previous successful attempts and this unsuccessful attempt are— • there was some egg yolk mixed with the egg whites; • the bowl was slightly wet.
3 **Make an inference or inferences**	• Egg whites will not stiffen when whisked if some egg yolk is mixed with them. • Egg whites will not stiffen when whisked in a wet bowl. • Egg whites will not stiffen when whisked if both of the above conditions exist.
4 **Test your inference or inferences**	To test inference 1, crack an egg, separating the white from the yolk but leaving some yolk in it. Whisk the mixture in a dry bowl and record your results. To test inference 2, crack an egg separating the white from the yolk. Place the white in a damp bowl. Whisk the mixture and record your results. To test inference 3, crack an egg, separating the white from the yolk but leavin some yolk in it. Place the white in a damp bowl. Whisk the mixture and record your results. As a control, crack an egg, separating the white from the yolk. Place the white a dry bowl and whisk. Compare the results with the control and write a conclusion.

You will use the same process to solve the problem at the end of this chapter.

Chapter 4 – Explaining reactions 978 1 4202 2997 4

The section of the workbook called During reading contains exercises that ask you questions about the ideas on certain pages of your textbook.

The opening section of the first 10 chapters is called Problem solving. It contains a problem-solving strategy for you to work through. You will apply this strategy to a problem in the Roundup on the last page of the chapter.

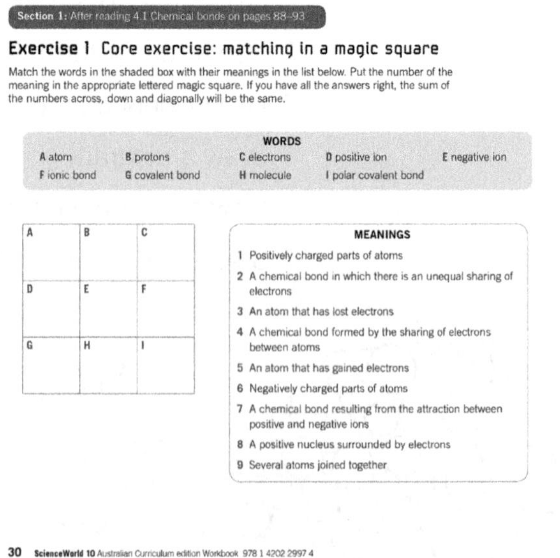

During reading

In this chapter, there are three main sections in the textbook:

4.1 Chemical bonds (pages 88–93)

4.2 Chemical shorthand (pages 95–100)

4.3 Predicting a reaction (pages 104–107)

For each of these sections, you will do Core exercises to ensure that you have a basic knowledge and understanding of the information in the textbook, and then do Challenge exercises to use and apply your knowledge and understanding.

Section 1: After reading 4.1 Chemical bonds on pages 88–93

Exercise 1 Core exercise: matching in a magic square

Match the words in the shaded box with their meanings in the list below. Put the number of the meaning in the appropriate lettered magic square. If you have all the answers right, the sum of the numbers across, down and diagonally will be the same.

WORDS

A atom, B protons, C electrons, D positive ion, E negative ion, F ionic bond, G covalent bond, H molecule, I polar covalent bond

A	B	C
D	E	F
G	H	I

MEANINGS

1 Positively charged parts of atoms

2 A chemical bond in which there is an unequal sharing of electrons

3 An atom that has lost electrons

4 A chemical bond formed by the sharing of electrons between atoms

5 An atom that has gained electrons

6 Negatively charged parts of atoms

7 A chemical bond resulting from the attraction between positive and negative ions

8 A positive nucleus surrounded by electrons

9 Several atoms joined together

30 ScienceWorld 10 Australian Curriculum edition Workbook 978 1 4202 2997 4

The Exercises contain questions for you to answer. Sometimes they require you to write words in the space provided. Others ask you to draw data tables or graphs. Some ask you to create stories or to solve problems.

Roundup contains a problem to be solved by applying your understanding of the information in the chapter and the problem-solving strategies you have learnt in this workbook.

Chapter 1

Investigating reactions

Problem solving

Problem solving strategy:

Break down the problem into manageable parts and then work through them one at a time.

Many scientific problems cannot be solved in one easy step. You need to work through the problem one step at a time, starting from what you know, then developing a possible solution, testing it, and analysing the results. If the test is fair, the results support your hypothesis, and repeating the test gives the same results, then you have a solution. If not, you need to start again.

Situation: Every time Sam goes into the garden in spring, she starts sneezing uncontrollably. This does not happen in any other season, and never occurs inside the house.

Problem-solving process	Problem-solving example
1 **Planning**	
Identify the problem	What is in the garden in spring only that makes Sam sneeze?
Identify the variables.	The only variable in spring is that many plants are in flower.
Write a possible hypothesis.	One or more of the flowering plants is making Sam sneeze.
Work out the variables you will change, measure and control.	Test each flowering plant one at a time under the same conditions.
Work out a method and select equipment.	Sam sniffs each plant one at a time inside the house (so that she can't smell other flowers in the garden). Apart from a selection of flowers, no equipment is needed.
2 **Conducting the investigation**	
Carry out the experiment. Observe, measure and record data.	Sam sniffs each flowering plant one at a time and records the effect.

3 Processing the data Organise the data in a table, and if necessary, plot a graph and do calculations.	**Plant** wattle clover bottlebrush orange blossom rose	**Sam's reaction** sneezed no effect sneezed no effect no effect
Identify any patterns and relationships between the variables.	Sam's sneezing is caused by wattle and bottlebrush flowers.	
Use your scientific knowledge to explain the patterns and relationships.	Hay fever can be caused by flowering plants.	
4 Evaluating the investigation Are the results valid? Is the conclusion valid?	Conduct the test again (or maybe several times) at a different time on another day. If the results agree with your original ones then they are valid and you can draw a valid conclusion.	
Did the results answer the original problem?	Yes, the findings relate directly to Sam's sneezing in spring.	

You will use the same process to solve the problem at the end of this chapter.

What do you know already?

Exercise 1 Inferring

1 A block of wood was placed in a sealed non-flammable container. The mass of the container plus the wood was 2 kg.

The wood was then burnt inside the sealed container and, after burning, the mass was still 2 kg.

When the container was opened, however, the mass of the container and the remains of the wood was less than it was to start with. The experiment was repeated several times, always with the same result.

Suggest a possible explanation for the decrease in mass.

During reading

Exercise 2 Explaining

After reading pages 3 and 6

Complete the following cartoons.

Exercise 3 Generalising and hypothesising

After doing Investigation 1 on pages 4–5

A hypothesis is a generalisation which can be tested.

Generalisations can be written in three different ways:

- As a statement— e.g. Paint dries more quickly on warm days.
- As a comparison— e.g. The warmer the day, the faster paint dries.
- Using *Generalisation linking words* such as those found on page 99 of this workbook. e.g. Paint usually dries more quickly on warm days.

Tick the generalisations below that can be scientifically tested.

1 The higher the temperature, the faster the reaction.

2 Most people prefer jazz to rock music.

3 Martians are more intelligent than all other forms of life.

4 Science teachers are usually better looking than other teachers.

5 Becky can lift heavier weights than David.

In a group, discuss your answers above and describe how you would go about testing the hypotheses you ticked.

Exercise 4 Creating and interpreting a line graph

After doing Investigation 1 on pages 4–5

1 A graph is an excellent way of showing the relationship between two measurements. Use the measurements in the data table to plot a graph, then draw a smooth curve through the points.

The data in the table shows the average growth of a number of bean seedlings kept at different temperatures. All seeds were grown in identical conditions, and only the temperature was varied.

Temperature (°C)	Average plant growth (cm)
0	1
5	1.5
10	3
20	4
40	5.5

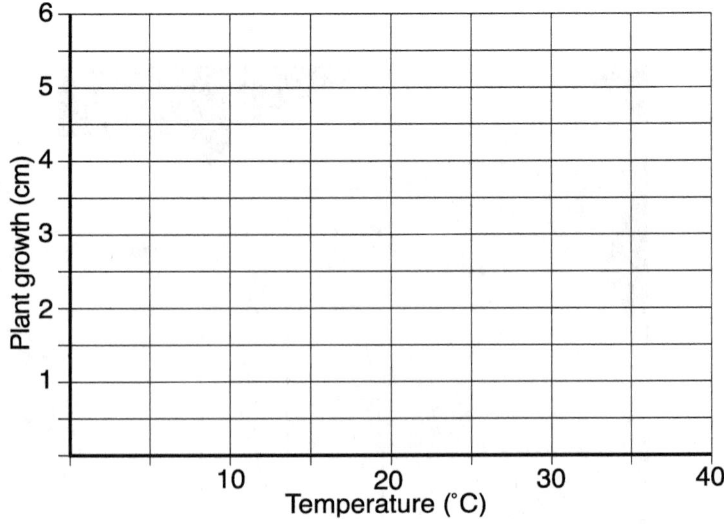

2 Write a hypothesis which generalises about the information above ______

Exercise 5 Formulating a hypothesis from observations

After reading Concentration and Surface area on page 7

Complete the hypothesis.

Exercise 6 Linking causes and effects

After reading Speeding up reactions on page 11 and doing Investigation 2 on pages 12–13

Complete the causes for the effects on the right.

	Cause		Effect
1	Increasing the ____________________	→	faster reaction rate
	Increasing the ______________ of the reactants	→	faster reaction rate
	Increasing the ______________ of the reactants	→	faster reaction rate
	Using a ________________________	→	faster reaction rate
2	Decreasing the ____________________	→	slower reaction rate
	Decreasing the ______________ of the reactants	→	slower reaction rate
	Decreasing the ______________ of the reactants	→	slower reaction rate
	Using an ________________________	→	slower reaction rate

Exercise 7 Using cause and effect linking words

The arrows used in Exercise 6 can be changed into *Cause and effect linking words*.

For example, instead of—

Increasing the temperature → faster reaction rate.

you would write the sentence—

Increasing the temperature *results in* a faster reaction rate.

or

A faster reaction rate *is due to* increasing the temperature.

Use *Cause and effect linking words* from page 99 of this workbook to substitute for the arrows in these word equations.

1 Decreasing the surface area of the reactants → slower reaction rate.

____

2 Using an inhibitor → slower reaction rate.

____

3 Increasing the concentration of the reactants → faster reaction rate.

____

Exercise 8 Contrasting

After reading Exothermic and endothermic on page 17

1 Sketch a diagram to illustrate the difference between an exothermic reaction and an endothermic reaction. Write captions under your diagrams to explain them.

An exothermic reaction is ______________________________

An endothermic reaction is ______________________________

2 When you are asked to contrast information (describe the differences between), some useful linking words are *although*, *however* and *on the other hand*. See the list of *Contrast linking words* on page 99 of this workbook.

Example: A dilute solution is one which has a small amount of substance dissolved in a liquid, *while* a concentrated solution has a larger amount of substance dissolved in it.

Write a sentence to contrast an exothermic reaction and an endothermic reaction. ______________________________

____

____

Exercise 9 Observing, inferring, predicting and hypothesising

After reading Lavoisier and combustion on page 19

The French scientist Antoine Lavoisier set out to solve the problem of what happens to the mass of the reactants and products during combustion.

1 a When he heated tin in a sealed flask, what was his observation about the mass? ______________________

____

b When he opened the flask, what was his observation about the mass? ______________________

____

c What was his inference from these two observations? ______________________

____

d What was his prediction? ______________________

____

e What was his hypothesis? ______________________

____

f After all his experiments, what was his conclusion? ______________________

____

2 Lavoisier's hypothesis has always been proven correct, so it is now called the *law of conservation of mass*. Copy this law from the textbook (page 21) and then explain what it means in your own words.

Law of conversation of mass	
Textbook language:	*Your explanation:*

Exercise 10 Completing word equations

Complete the following equations for these reactions.

1 sodium thiosulfate + [ ] → sulfur + other products

2 [ ] + oxygen → tin oxide

3 [ ] + water + energy → plant food + [ ]

4 hydrogen peroxide → [ ] + oxygen

Exercise 11 Finding key words

1 Use the clues below to find the answers in the box. (The answers may be written horizontally or vertically, backwards or forwards.)

2 When you find the answer, cross out those letters. (Some letters may be used more than once.)

3 When you have crossed out the answers to all the clues, 14 letters will remain. Use these letters to form another explosive word used in this chapter.

C	I	M	R	E	H	T	O	D	N	E	N	I
I	C	O	L	L	O	I	D	U	R	E	A	C
M	L	T	R	N	O	I	T	C	A	E	R	O
R	O	P	Y	H	I	E	M	Y	Z	N	E	N
E	T	N	E	D	N	E	P	E	D	N	I	C
H	T	O	C	G	H	S	A	S	I	S	S	E
T	I	A	A	L	I	U	R	A	L	O	I	N
O	N	E	T	Y	B	R	T	L	U	D	O	T
X	G	R	A	R	I	F	I	Y	T	I	V	R
E	S	A	L	A	T	A	C	M	E	U	A	A
C	E	C	Y	T	O	C	L	A	R	M	L	T
I	N	E	S	E	R	E	E	A	C	I	D	E
N	O	I	T	S	U	B	M	O	C	O	L	D

Clues

1 A reaction in which some form of energy must be supplied.

2 A reaction in which energy is released.

3 Another word for burning.

4 A substance that speeds up a reaction.

5 A solution that contains only a small amount of solute.

6 A French scientist whose experiments disproved the phlogiston theory.

7 The enzyme responsible for decomposition of hydrogen peroxide in the body.

8 A slowing-down catalyst.

9 A biological catalyst.

10 An action that occurs as the result of another action.

11 Another name for sodium thiosulfate.

12 In Investigation 1, Part B, the mixture was cloudy due to the sulfur ________.

13 The variable that is purposely changed.

14 The ________ theory of matter is used to explain reaction rates.

15 Enzyme that breaks down starch.

16 A solution that contains a large amount of solute.

17 Another word for speed.

18 The exterior face of an object.

19 The extent of a 2-D surface.

20 Baking soda is the name for ________ hydrogen carbonate.

21 Vinegar is the name for acetic ________.

22 Is a reaction slower in heat or cold?

23 The process in which rennin reacts with milk and solidifies it.

24 The inhibitor that makes nitroglycerine explosives safer.

25 Where would you find a catalytic converter?

Letters remaining

Roundup

Exercise 12 Problem solving using the strategy:

Break down the problem into manageable parts and then work through them one at a time.

Jessica grew 20 tomato plants from seeds. She planted 10 tomato plants in a garden on the eastern side of her house. Then she planted 10 plants in a garden on the western side.

The tomato plants in the eastern garden grew taller and greener and produced more fruit than the western garden plants. How would Jessica work out why this happened?

Use the checklist on the left hand side of the table as a guide to solve this problem

You may have to write this in your notebook.

Problem-solving process	Your solution
1 **Planning** Identify the problem. Identify the variables. Write a possible hypothesis. Work out the variables you will change, measure and control. Work out a method and select equipment.	__________ __________ __________ __________ __________
2 **Conducting the investigation** Carry out the experiment. Observe, measure and record data.	You do not need to actually conduct the experiment, but you do need to show how you would record the results.
3 **Processing the data** Organise the data in a table, and if necessary, plot a graph and do calculations. Identify the patterns and the relationships between the variables. Use scientific knowledge to explain patterns and relationships.	__________ __________ __________ __________ __________
4 **Evaluating the investigation** Are the results valid? Is the conclusion valid? Did the results answer the original problem?	__________ __________ __________

Chapter 2

Road science

Problem solving

Problem solving strategy:

Devise a table, chart or graph.

When you do experiments or collect relevant data to solve problems, you need to organise your results or data in a way that clearly compares them. The best way of doing this is by devising a table, chart or graph.

Situation: Two radio-controlled cars cross the starting line together. A datalogger is used to record the position of each car for the first 10 seconds of the race. How can the data be used to find out which car was in front after 5 seconds, and whether one car overtook the other during the first 10 seconds, and, if so, when?

Car A

Time (s)	Dist from start (m)
0	0
2	8
4	16
6	24
8	32
10	40

Car B

Time (s)	Dist from start (m)
0	0
2	2
4	8
6	16
8	29
10	50

Problem-solving process	Problem-solving example
1 **Identify the problem**	a Which car, A or B, was in front after the first 5 seconds? b Did one car overtake the other during the first 10 seconds? If so, when?
2 **Devise a table, chart or graph** Gather the data you need to solve the problem and present it in a table, chart or graph so that you can see any patterns in the data.	You can answer both parts of this problem by drawing a graph of distance versus time for both cars. 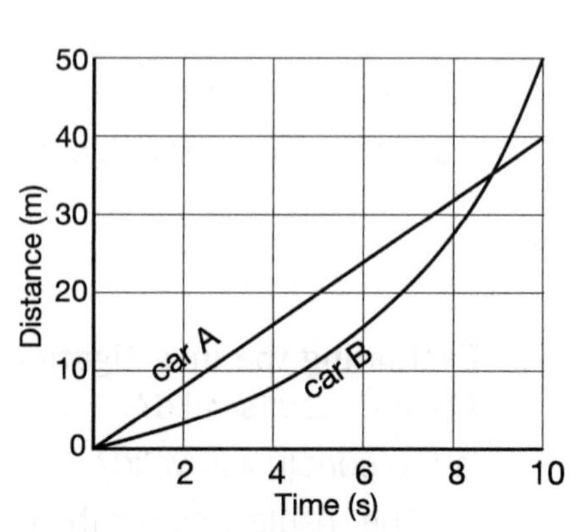

3 **Interpret your data to solve the problem**	From the graph you can see that after 5 seconds Car A has travelled 20 m, while Car B has travelled 12 m. The graph also tells you that to start with Car A was in front, but at about 8.8 seconds Car B passed Car A. You know this because the graphs cross.

You will use the same process to solve the problem at the end of this chapter.

During reading

During this chapter you will see stars next to each task.

★ means a simple task which you should be able to do quickly using information from the textbook.

★★ means a task which will require you to infer, interpret, or reach a conclusion from the information in the textbook.

★★★ means a more difficult task which will require you to generalise or apply the information in the textbook more widely.

Exercise 1 Using a mathematical formula

After reading Speed on pages 29–30

★ **1** Calculate the average speed of a car which travelled 200 kilometres in 4 hours. Fill in the boxes.

$$\text{average speed} = \frac{\text{distance}}{\square} = \frac{\square\ \text{km}}{\square\ \text{h}} = \square\ \text{km/h}$$

★ **2** Calculate the distance travelled by a bus in 3 hours at an average speed of 65 km/h.

$$\text{distance} = \square \times \square = \square \times \square = \square\ \text{km}$$

★★ **3** A cyclist rides 90 km at an average speed of 40 km/h. How long does she take—in hours and minutes? (Show your working, including the formula, as in **1** and **2**.)

★★ 4 Which of the following is the fastest speed? Show your working.

a 60 km/h

b 15 m/s

c 1.1 km/min

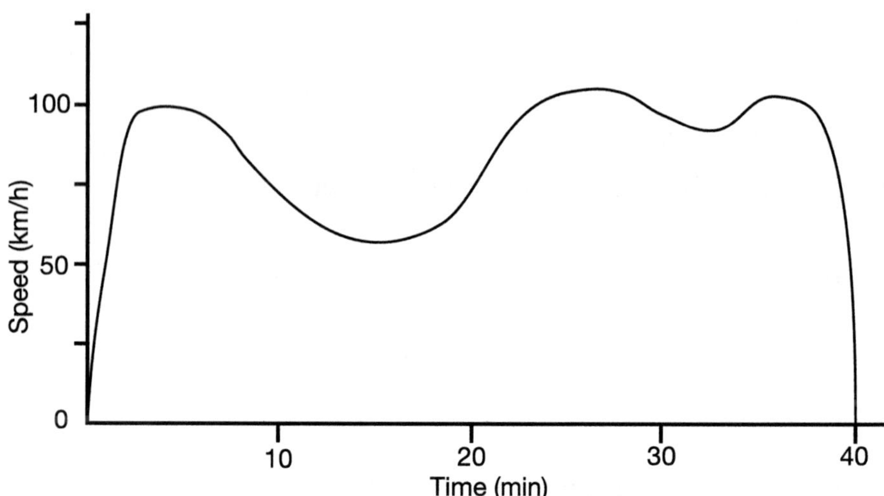

★★★ 5 On the weekend James drove a distance of 58 kilometres to the beach. The graph shows his speed for the trip.

a What was his highest *instantaneous* speed during the trip? ______________________

b Suggest two reasons for the dip in the graph between the 7th and 23rd minute of the trip.

____

____

____

c Use the graph (not the formula) to estimate his average speed for the trip. ______________________

d Now use the formula to calculate his average speed. Compare this result with your estimation in **c**.

Exercise 2 Calculating acceleration

After reading Acceleration on page 31 and doing Investigation 5 on pages 32–34

★ **1** Calculate the acceleration of a horse which gallops from rest to 20 m/s in 4 seconds.

average acceleration = $\dfrac{\square - \text{initial speed}}{\square}$

= $\dfrac{\square - \square}{\square}$

= $\square$ m/s²

★ **2** The same horse takes 5 seconds to stop when galloping at 30 m/s. What is its deceleration?

deceleration = $\dfrac{\square - \square}{\square}$

= $\square$ m/s²

★ **3** Your answer for **2** should have been a negative number. Why is this so?

____

★★ **4** The Porsche 911 accelerated at 9.3 m/s². Explain in your own words exactly what m/s² means.

____

____

★★ **5** Below is a ticker tape which was attached to a slot-car moving at a variable speed. Mark on the ticker tape where the car was—

- accelerating (positive acceleration)
- decelerating (negative acceleration)
- moving at a constant speed (zero acceleration)

You may want to use different coloured pens.

START

|•• • • • • • • ••••• • • • • • • • • • • • •••

★★★ 6 Complete the graph below showing how the acceleration of the slot-car above varies.

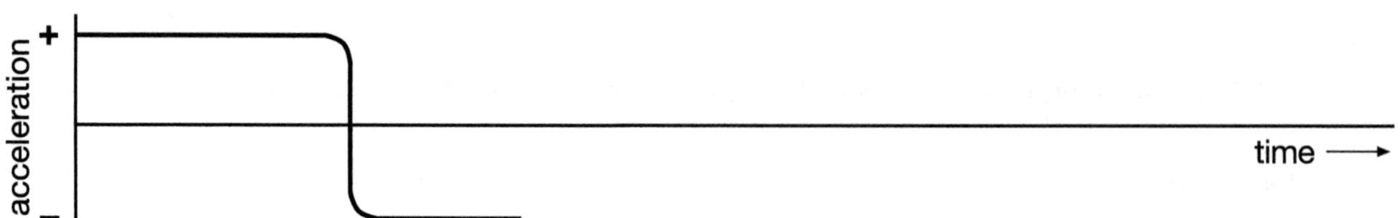

★★★ 7 Jade knows that if you drop something it accelerates as it falls to the ground. She measures the distance from the ground to the top of a building (9.7 m), then times how long a stone takes to fall (1.4 s). Use the following formula to help Jade calculate the acceleration due to gravity:

$$d = ut + \tfrac{1}{2} gt^2$$

where d = distance

u = initial speed

t = time

g = acceleration due to gravity

Hint: Since the stone starts from rest, its initial speed is zero.

Exercise 3 Critical reading

After reading Stopping on page 38

★★ 1 Use the information on page 38 to complete the following paragraph.

It is a fine day and you are travelling at 100 km/h (27.8 m/s) when you see a stationary car 100 metres in front of you. It takes 1 second for you to press the brake pedal. This is called your ________________. During this time you travel ____________metres. It is another ____________ metres before the vehicle comes to a halt. This is called your ______________________. You have therefore travelled a total of ____________ metres since you first saw the car. This is called your______________________. If it had been raining your reaction distance would have been ____________________, but your braking distance would have been ____________ metres, giving a stopping distance of ______________ metres. The result of this would be that ____.

★★ 2 Complete the teacher's blackboard summary about the variables that affect the stopping distance of a car.

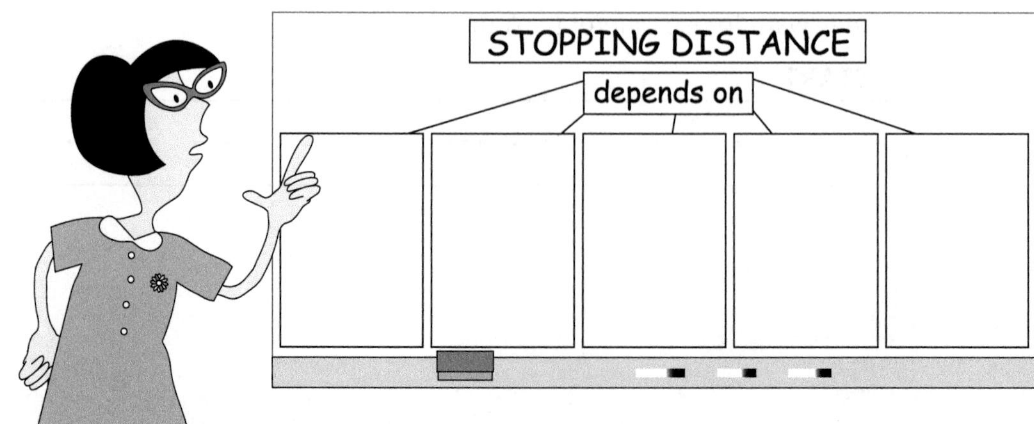

Exercise 4 Recalling information

After reading Inertia on page 44

★ **1** Decide whether each of the following statements is true or false.

a The heavier the person, the less inertia that person's body has. TRUE FALSE

b A heavy object needs a larger force to stop it than a lighter object moving at the same speed does. TRUE FALSE

c A 120 kg footballer running towards the try line has more momentum than an 80 kg footballer running at the same speed. TRUE FALSE

d A 50 kg girl running at 15 km/h has less momentum than a 60 kg boy running at 12 km/h. TRUE FALSE

★ **2** Complete the caption for the cartoon below, using the correct scientific terms.

The limousine is heavier than my sports car. In other words, it has more ______________ Because of this, it has a greater ______________, and requires a larger ______________ to get moving.

★★ **3** Write a caption for this cartoon using the word *inertia*.

____

____

____

____

____

____

Exercise 5 Interpreting tables and graphs

After reading Inertia and car design on pages 47–48

A survey was done to find out how successful seatbelts are in reducing injuries in car accidents. Use the results to answer the questions on the following page.

Injuries to unbelted drivers

Type of injury	Speed (kilometres per hour)								
	not moving	less than 20	20–40	about 50	about 60	70–90	about 100	more than 100	Total
Fatal injuries	0	2	5	2	1	13	5	9	37
Serious injuries	7	12	50	44	45	74	26	5	263
Light injuries	35	65	245	126	115	189	40	20	835
Non-injured	2302	4865	7643	2446	1427	1462	175	51	20 371
Total	2344	4944	7943	2618	1588	1738	246	85	21 506
Percentage of total accidents	11	23	37	12	7	8	1	1	100

Injuries to belted drivers

Type of injury	Speed (kilometres per hour)								
	not moving	less than 20	20–40	about 50	about 60	70–90	about 100	more than 100	Total
Fatal injuries	0	0	0	0	0	0	3	3	6
Serious injuries	9	3	19	25	16	60	16	13	161
Light injuries	47	41	106	106	72	125	38	13	548
Non-injured	2266	4229	6786	2661	2053	2398	297	100	20 790
Total	2322	4273	6911	2792	2141	2583	354	129	21 505
Percentage of total accidents	11	20	32	13	9	12	2	1	100

★★ 1 From the information in the two tables, draw a table in the box below that compares the total number of injuries to unbelted and belted drivers for all speeds. You will need to work out how many rows and columns the table will need, and which headings to use.

Note: Total number of injuries = fatal + serious + light injuries

★★ 2 Use the table you have drawn up to complete the bar graph below. The first bars are done for you, but you will need to add labels to show which are belted and which are unbelted.

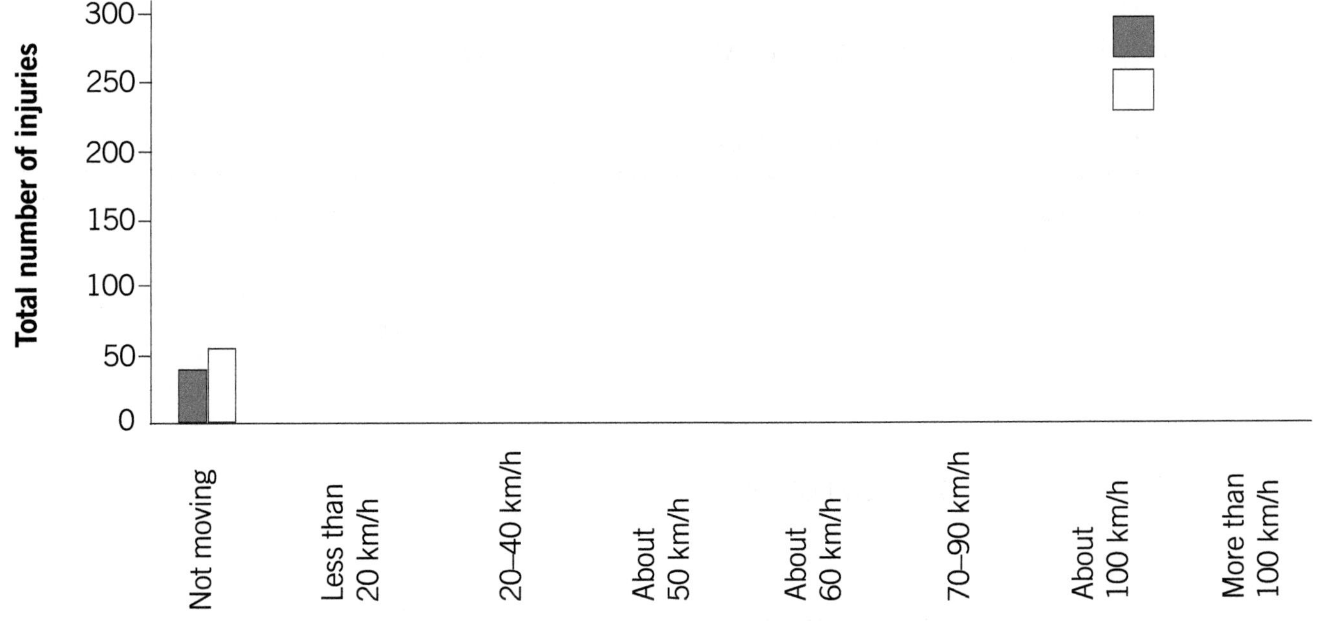

★ 3 At which speed did most injuries occur? ______________________

★ 4 At which speed was the difference between belted and unbelted injuries greatest? ______________________

★ 5 Write a generalisation linking total number of injuries to whether or not the drivers were wearing a seat belt.

____

____

★★ 6 Is there an exception to this generalisation? Explain. ______________________

____

Exercise 6 Listing in sequence

After reading Inertia and car design on pages 47–48

★★ The events listed below occur when a car fitted with a driver's airbag is involved in a major collision. Rearrange the events in the correct order by numbering the boxes.

- airbag deflates ☐
- airbag inflates ☐
- airbag cushions driver's body ☐
- car collides with other vehicle ☐
- chemicals in inflator module are mixed ☐
- driver's inertia causes her to continue moving forward ☐
- electric current flows in inflator module circuit ☐
- nitrogen gas is produced ☐

Exercise 7 Constructing an overview

After reading pages 44–49

Use the words in the box below to complete an overview of section 2.3 Collisions.

- seat belts
- **m** x **a**
- force
- head restraints
- rigid passenger compartment
- air bags
- inertia
- crumple zone
- **m** x **v**

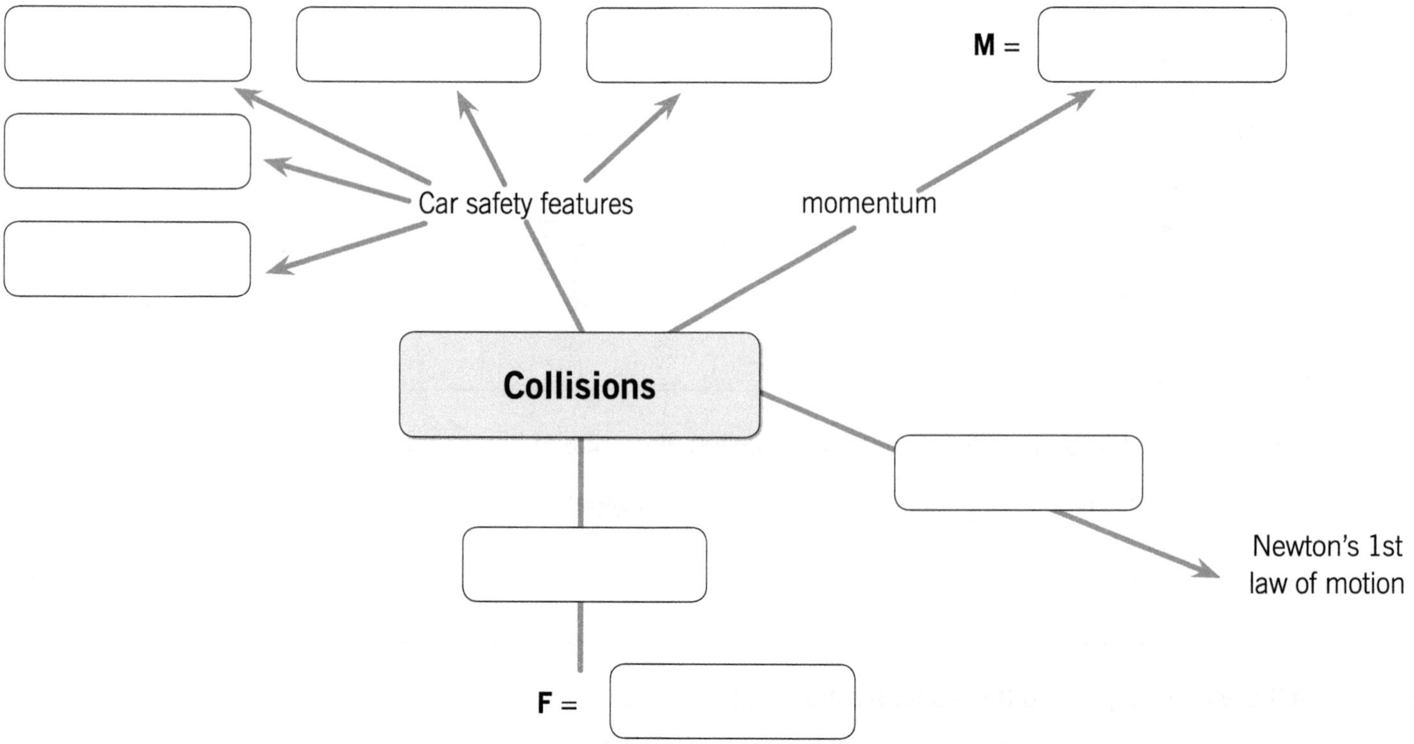

Car safety features

momentum

M =

Collisions

Newton's 1st law of motion

F =

Roundup

Exercise 8 Problem solving using the strategy:

Devise a table, chart or graph.

Ella has her own radio-controlled car which she races almost every weekend. She has been invited to compete in a race on a track she has not seen. Her friend, Matt, has raced his car on the track, and tells her it has six sections. It starts with a straight flat section, but he cannot remember the order of the other sections. He thinks that after the first straight flat section comes a short rough section followed by a tight corner. He also knows there is a hill, another straight section and another tight corner.

To help Ella with her preparations, Matt has a record of the speed of his car over the 10 seconds it took to complete the course. Here is his data—

The car started from rest and reached 12 m/s after 1 second; it dropped speed to 8 m/s at 2 seconds; then at the 6th second it was travelling at 10 m/s. At 7 seconds its speed was 8 m/s and then at 8 seconds it was only 2 m/s. By the end of the race, at the 10 second mark, it was travelling at 12 m/s.

Use this information to help Ella work out the layout of the track she has not seen.

Problem-solving process	Your solution
1 **Identify the problem** Write the problem in your own words.	__________ __________ __________ __________
2 **Devise a table, chart or graph** Gather the data you need to solve the problem and present it in a table, chart or graph so that you can see any patterns in the data.	
3 **Interpret your data to solve the problem**	__________ __________ __________

Chapter 3
Inheritance

Problem solving

Problem solving strategy:

Work backwards to eliminate other possibilities

Some problems have several possible solutions. In this case, you need to list all the possibilities and then work backwards to find the correct solution.

Situation: Person X has been recently diagnosed with HIV (AIDS). X wants to identify how the virus was transmitted to him.

Problem-solving process	Problem-solving example
1 **Identify the problem**	How was the virus transmitted to X?
2 **Using your scientific knowledge, identify all the possibilities**	X knows that HIV (the virus that causes AIDS) is spread in the blood mainly by sharing needles (taking drugs intravenously). It can also be spread by contact with infected blood (e.g. blood transfusions, medical injections, and blood to blood contact), and in semen by sexual contact. X has not had a medical injection for 10 years and has never used drugs intravenously, so the possibility that the virus has been transmitted by sharing needles is immediately eliminated. X was tested for HIV in 2002 and was negative. He came into contact with blood from person Z when they were both involved in a car accident in 2004. X has been married to Y since 1995. Y also has been recently diagnosed with HIV, but was tested for HIV in 2002 and was negative. Y had a blood transfusion after the birth of her baby in 2001. Y has had no medical injections since 2002.

3 Work backwards
Using time as your organising structure, work backwards from the most recent events or factors to the earliest events or factors.

Person	Year	Contact	HIV test
X	2004	blood contact with Z	+ve
	2004–1995	sexual contact with Y	–ve
Y	2004–2002	sexual contact with X	+ve
	2002–1995	sexual contact with X	–ve
	2001	blood transfusion	–ve

4 Eliminate all but one possibility
Work backwards until you have eliminated all other possibilities.

X knows that Y had a blood transfusion in 2001 but was HIV negative in 2002, so the HIV was not transmitted by the blood transfusion.

X infers that since Y has had no medical injection or blood to blood contact since 2002, the virus was not transmitted by her.

X infers that the source of the virus was person Z, and that the virus was then transmitted from him (X) to Y.

You will use the same process to solve the problem at the end of this chapter.

During reading

Exercise 1 Recalling information

After reading pages 58–61

Complete the Acrossword on the next page. Then rearrange the letters from the shaded column to create another key word. The answer is on page 23 of this workbook.

1 Qualities that distinguish one organism from another
2 Sausage-shaped objects that can be seen in cells just prior to division
3 Any group of individuals born about the same time, e.g. parents are one group, children are another
4 The X and Y chromosomes determine the __________ of an individual.
5 Plural of nucleus
6 The organism that has 23 pairs of chromosomes
7 The male sex cell
8 American biologist who studied inheritance
9 The process of cell division
10 Features that are passed down from one generation to the next
11 An organism that has eight chromosomes (2 words)
12 The building blocks of all living things
13 The female sex cell

1
2
3
4
5
6
7
8
9
10
11
12
13

Shaded letters = __ __ __ __ __ __ __ __ __ __ __ __ __

Key word = ______________________________

Exercise 2 Constructing an overview

After reading Determining sex on page 61

Use the information in the textbook to decide which of these two statements is correct. Justify your decision with scientific facts.

____

____

____

____

Exercise 3 Completing an overview

After reading pages 64–66

Use the words in the box below to complete an overview of section 3.2 DNA.

- deoxyribose
- thymine (T)
- one amino acid
- DNA
- adenine (A)
- cytosine (C)
- proteins
- chromosomes
- bases (N-containing)
- phosphates

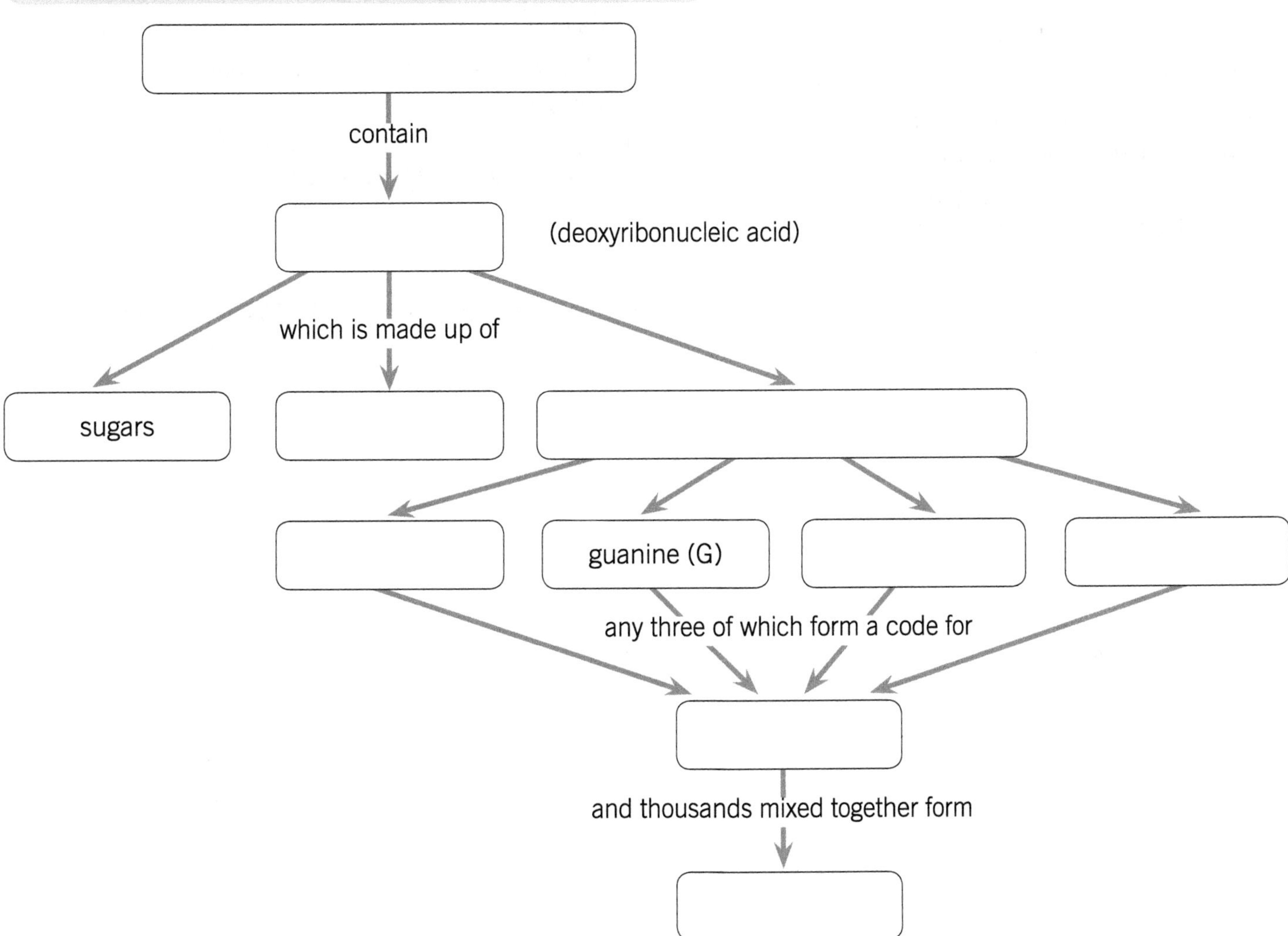

Answer

Exercise 1

Fertilisation

Exercise 4 Using cause and effect linking words

After reading pages 66–68

Use *Cause and effect linking words* from page 99 of this workbook to complete the following summary. Try to use a variety of linking words and read over your summary carefully to make sure it is correct.

________________ mutations can be either harmful or beneficial. Mutations ________________ alterations to genes. Mutations ________________ changes to characteristics through changes in part of the DNA strand, ________________ the production of a quite different protein. Mutations can occur spontaneously or can be ________________ overexposure to the sun, high-energy radiation or chemicals.

Exercise 5 Expressing points of view

The outcomes of the Human Genome Project have raised arguments for and against DNA testing.

1 Choose a role from those listed below (or create your own) and prepare an argument FOR or AGAINST these scientific advances. Share this point of view with the rest of the class.

- A person wrongly accused of a crime
- An employer who wants access to genetic information before employing staff
- A detective investigating rape crimes
- The Chairman of a Health Insurance Company
- A man or woman whose unborn baby has been diagnosed with Down syndrome
- An Australian scientist whose research contributed to the Human Genome Project
- A young man who is unsure whether his girlfriend is pregnant to him or someone else
- A very poor Indian person living in the slums of Calcutta
- The chairperson of a large multinational drug company
- A parent whose child disappeared and was probably killed 20 years ago

2 When you have listened to these role-play arguments, break into small groups and share your *own* point of view.

Working space

Exercise 6 Defining, describing and predicting

After reading pages 73–78

1 Complete the definitions for each of the following terms.

a A dominant gene is a gene that ______________________________

______________________________. An example is ______________.

It is usually represented by a capital letter.

b A recessive gene is a gene that ______________________________

______________________________. An example is ______________.

It is usually represented by a lower case letter.

c A homozygous or pure breeder is an organism in which ______________________________

______________________________.

d A heterozygous or hybrid breeder is an organism in which ______________________________

____

e Incomplete dominance is a genetic situation in which ______________________________

______________________________. An example is ______________.

f Co-dominance is a genetic situation in which ______________________________

______________________________. An example is ______________.

2 Complete the following descriptions.

The type of genes in an organism is called its ______________, whereas ______________

______________ is called its phenotype.

3 a If a smooth-seed pea with genes Ss is crossed with a wrinkle-seed pea with genes ss, predict the ratio of smooth-seed peas to wrinkle-seed peas in the next generation. Use the Punnett square on the right to work it out.

Ratio __________ : __________

b If the Punnett square for the first generation of fruit flies is as shown below, where R represents the dominant gene for red eyes, what prediction can you make about—

	R	r
R	Rr	Rr
r	Rr	

- the parents?

____

____

- the next generation if one of these Rr flies mates with a fruit fly with rr genes?

____

____

Exercise 7 Writing in role

After reading page 79

Respond to this letter.

Dear Geneticist,

I have blood type A and would really like to have children with blood type AB. Which blood types should I look for in potential brides?

Dracula

Draw a pedigree or a Punnett square.

Dear Dracula,

____

____

____

____

____

____

____

____

Yours sincerely

____

Exercise 8 Writing a play scene

After reading page 80

Complete the following dialogue, with stage directions, between a geneticist and her patient. You will need to do some research to find information on haemophilia. When you have completed the script, rehearse in pairs to present it to your class.

Setting: The geneticist's rooms.
The two characters are sitting on chairs facing each other.

Characters: The geneticist, a caring person with a wide knowledge of inherited conditions.
The patient, a frightened 14-year-old boy who has just been told he has a mild form of haemophilia.

Patient: (*in a frightened voice*) But ... but ... what is haemophilia?

Geneticist: (________________) ________________________________

____

____

Patient: (angrily) How do you know that's what I've got? Can it be cured?

Geneticist: (________________) ________________________________

____

____

Patient: (*still angry*) Well, how did I get it? My sister hasn't got it!

Geneticist: (________________) ________________________________

____

____

Patient: (*calming down*) So my mother is a carrier but she hasn't got it?

Geneticist: Exactly!

Patient: How come?

Geneticist: Well, it's all to do with X-linked genes.

Patient: (________________) What are X-linked genes?

Geneticist: ____

____

____

Patient: (*sadly*) Well, how will this affect my life?

Geneticist: (________________) ________________________________

____

____

Roundup

Exercise 9 Problem solving using the strategy:

Work backwards to eliminate other possibilities.

In the pedigree below, the grey shaded squares show males with haemophilia, and the grey shaded circles show females with haemophilia. (Haemophilia is a disease in which the blood does not clot. The genes that cause haemophilia are X-linked.)

The condition of the people represented by black squares and circles is unknown. Use the pedigree to work out whether these people have haemophilia or carry the gene.

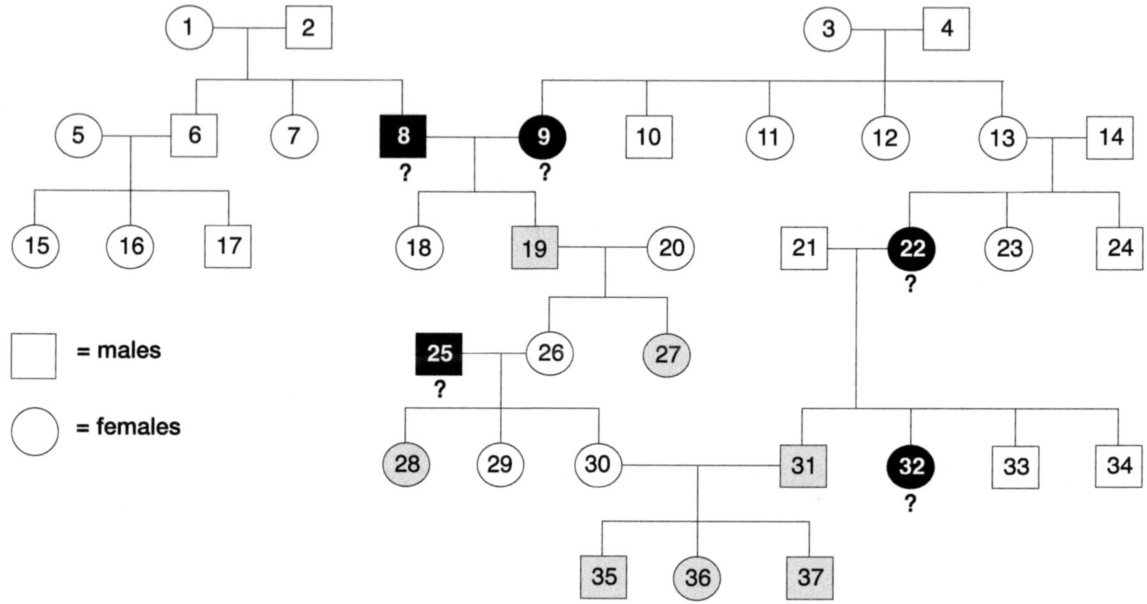

Note: the steps in the problem-solving process below are slightly different from the one at the start of the chapter.

Problem-solving process	Your solution
1 **Identify the problem**	
2 **Work backwards** Using the pedigree as your organising structure, work backwards from the youngest generation to the earlier generations.	
3 **Using your scientific knowledge, identify all the possibilities** Work out the possible genotypes of the people in question.	

Chapter 4

Explaining reactions

Problem solving

Problem solving strategy:

Infer and test.

This strategy requires you to make observations, then to make an inference or inferences based on these observations, and finally to test your inference(s) scientifically.

Situation: When you make meringue, the first step is to whisk the whites of the eggs until the mixture is so stiff that you can turn the bowl upside down.

Imagine that you have done this successfully many times but that, on this occasion, the whites don't stiffen, no matter how much you whisk them.

The only differences you can recall are that there was a little egg yolk in the whites and that the bowl was slightly wet.

Problem-solving process	Problem-solving example
1 **Identify the problem**	Why won't the egg whites stiffen despite the whisking?
2 **List your observations**	In this case, the differences between previous successful attempts and this unsuccessful attempt are— • there was some egg yolk mixed with the egg whites; • the bowl was slightly wet.
3 **Make an inference or inferences**	• Egg whites will not stiffen when whisked if some egg yolk is mixed with them. • Egg whites will not stiffen when whisked in a wet bowl. • Egg whites will not stiffen when whisked if both of the above conditions exist.
4 **Test your inference or inferences**	To test inference 1, crack an egg, separating the white from the yolk but leaving some yolk in it. Whisk the mixture in a dry bowl and record your results. To test inference 2, crack an egg separating the white from the yolk. Place the white in a damp bowl. Whisk the mixture and record your results. To test inference 3, crack an egg, separating the white from the yolk but leaving some yolk in it. Place the white in a damp bowl. Whisk the mixture and record your results. As a control, crack an egg, separating the white from the yolk. Place the white in a dry bowl and whisk. Compare the results with the control and write a conclusion.

You will use the same process to solve the problem at the end of this chapter.

During reading

In this chapter, there are three main sections in the textbook:

4.1 Chemical bonds (pages 88–93)

4.2 Chemical shorthand (pages 95–100)

4.3 Predicting a reaction (pages 104–107)

For each of these sections, you will do Core exercises to ensure that you have a basic knowledge and understanding of the information in the textbook, and then do Challenge exercises to use and apply your knowledge and understanding.

Section 1: After reading 4.1 Chemical bonds on pages 88–93

Exercise 1 Core exercise: matching in a magic square

Match the words in the shaded box with their meanings in the list below. Put the number of the meaning in the appropriate lettered magic square. If you have all the answers right, the sum of the numbers across, down and diagonally will be the same.

WORDS

A atom | **B** protons | **C** electrons | **D** positive ion | **E** negative ion

F ionic bond | **G** covalent bond | **H** molecule | **I** polar covalent bond

A	B	C
D	E	F
G	H	I

MEANINGS

1 Positively charged parts of atoms

2 A chemical bond in which there is an unequal sharing of electrons

3 An atom that has lost electrons

4 A chemical bond formed by the sharing of electrons between atoms

5 An atom that has gained electrons

6 Negatively charged parts of atoms

7 A chemical bond resulting from the attraction between positive and negative ions

8 A positive nucleus surrounded by electrons

9 Several atoms joined together

Exercise 2 Core exercise: interpreting illustrations

1 Examine Fig 1 on page 88 carefully, then answer the questions below.

a What happens to the sodium atom to change it into a sodium ion? ______________________

____

b How many electrons are lost from the sodium atom? ______________________

c Does the sodium atom have a positive or negative charge or no charge? ______________________

d Does the sodium ion have a positive or negative charge or no charge? ______________________

2 Examine Fig 2 on page 88 which shows how a chlorine atom becomes a chloride ion.

a What happens to a chlorine atom to change it to a chloride ion? ______________________

____

b How many electrons are gained by the chlorine atom? ______________________

c Does the chlorine atom have a positive or negative charge or no charge? ______________________

d Does the chloride ion have a positive or negative charge or no charge? ______________________

Exercise 3 Challenge exercise: creating a table

Look at the four pairs of statements and the table below. In each pair there is a statement that belongs in the *Ionic bond* column of the table, and the other belongs in the *Covalent bond* column. Place each statement in the correct column.

1 { results from the attraction between positive and negative ions
formed by the sharing of electrons between atoms

2 { occurs when two non-metals react
occurs when a metal reacts with a non-metal

3 { substance does not form ions, so the solution does not conduct electricity
substance does form ions in solution, so will conduct electricity

4 { substance has a crystal lattice structure
substance consists of seperate molecules

Ionic bond	Covalent bond

Exercise 4 Challenge exercise: writing contrast sentences

The table you completed in Exercise 3 contrasts (shows the differences between) ionic bonds and covalent bonds.

Use *Contrast linking words* from page 99 of this workbook to join each pair of statements into one sentence. The first one is done for you as an example, using *while* as the linking word.

1 An ionic bond results from the attraction between positive and negative ions, *while* a covalent bond is formed by the sharing of electrons between atoms.

2 ______________________________

____

3 ______________________________

____

4 ______________________________

____

Exercise 5 Challenge exercise: explaining

After reading Ionic bonds on page 89

Complete the explanations in your own words.

Exercise 6 Challenge exercise: explaining

After reading Covalent bonds on pages 92–93

Complete the explanations in your own words.

OK! I understand how ions are formed. I understand that when a metal reacts with a non-metal, the positive and negative ions attract each other and so the compound is held together by an ionic bond. But what is a covalent bond?

Well, it's exactly the same principle except that ______________________________

... and a polar covalent bond occurs when ______________________________

Exercise 7 Core exercise: illustrating

Section 2: After reading 4.2 Chemical shorthand on pages 95–100

The formula for water is H_2O. Here is a diagram showing the atoms which make up a water molecule.

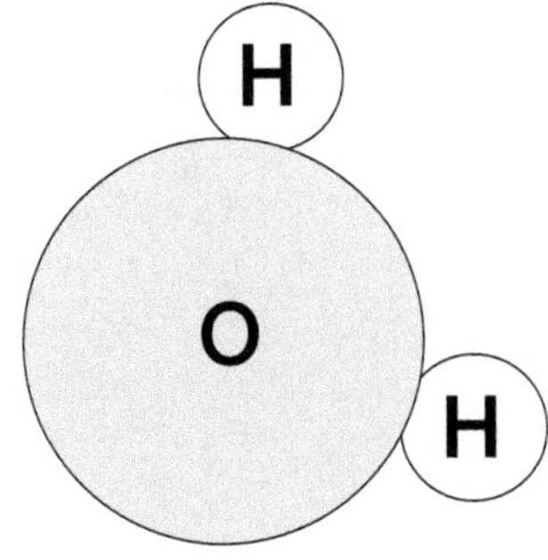

In the boxes below, draw and label diagrams showing the atoms which make up the molecules listed.

Oxygen molecule Formula: O_2	Methane molecule Formula: CH_4	Sulfur dioxide molecule Formula: SO_2

Exercise 8 Core exercise: interpreting diagrams

The boxes below contain diagrams of different molecules. Underneath each diagram is the name and formula of the molecule. Using the symbols in the formula, label each of the atoms in the drawing. The first one is done for you as an example.

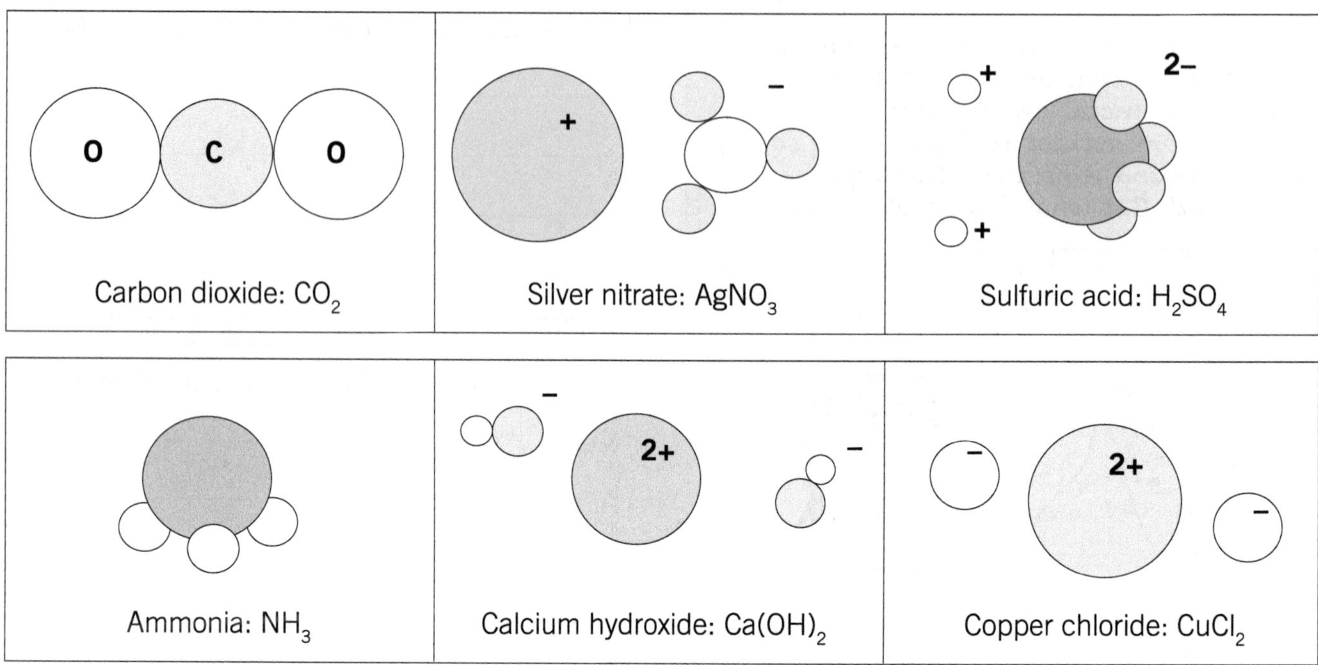

Exercise 9 Core exercise: interpreting formulas

For each formula, write down the names and numbers of each of the different atoms in the molecule. The first one is done for you.

$CFCl_3$ = 1 carbon, 1 fluorine and 3 chlorines

H_2SO_4 = ____________________

Na_2CO_3 = ____________________

$Zn(NO_3)_2$ = ____________________

$(NH_4)_2SO_4$ = ____________________

Exercise 10 Core exercise: interpreting a table

Closely examine the table on page 96 and answer the questions below.

1 What is the valency of aluminium? ______

2 What is the valency of the nitrate ion? ______

3 List the metals that have more than one valency. ______

4 Which negative ion has the highest valency? ______

5 If an ion has a combining power (valency) of 2–, what inference(s) can you make? ______

______

______

Exercise 11 Core exercise: writing formulas

Write formulas for the following ionic compounds.

1 silver nitrate: ______

2 copper (I) sulfate: ______

3 iron (II) chloride: ______

Exercise 12
Core exercise: writing formulas and balancing equations

Write the equations for what happens in each of the following common situations and then balance the equations.

1 Lactose ($C_{12}H_{22}O_{11}$), the sugar found in milk, reacts with oxygen to give carbon dioxide and water.

2 Aluminium reacts with oxygen to form a protective coating of aluminium oxide.

Exercise 13 Core exercise: writing in a role

The story below is a 'romantic' autobiography. As you read it, insert in the brackets [ ] the chemical symbols, formulas and reactions described. Use the information on page 96 if necessary.

Hello. My name is magnesium [ ] and I am always doubly positive [ ] about myself. Unfortunately, at the moment, I am going out with nitrate [ ] who is fairly negative [ ]. Together, we call ourselves magnesium nitrate [ ]. Our relationship is well balanced, but I have been looking out for another relationship. So I was thrilled when, one day as we were swimming, I saw sodium hydroxide [ ].

Of course I'm not interested in sodium [ ] as he is positive too [ ]. Actually, I'm quite repelled by sodium. But, wow, you should see hydroxide [ ]! Just gorgeous and, although negative like old nitrate, I think we could form a comfortable relationship. Hydroxide [ ] and I swam around for a while and then decided to stay together [+ →]. This was the solution to my relationship problem.

Much to my surprise, sodium and nitrate seemed to be getting along very well too. They also decided to stay together [+ →]. Now all four of us are happy and well balanced in our relationships.

Exercise 14 Challenge exercise: predicting

In words, not symbols, predict which substances may be formed when the following substances are mixed.

1 hydrogen + chloride →

2 calcium carbonate + sodium hydroxide →

3 potassium sulfate + lead nitrate →

Roundup

Exercise 15 Problem solving using the strategy:

Infer and test.

Situation: Bruno has a large vegetable garden. He grows his vegetables organically and uses water which comes from a spring near the garden to irrigate them. The spring is on the side of a hill which is made of both limestone and basalt rocks. Iron and manganese ions are often found in water that flows through basaltic soil.

This year Bruno had a bumper crop of carrots, so he decided to preserve the ones he could not eat or sell. He used the spring water to soak and clean the carrots. However, after soaking them overnight, he was surprised to see that they were soft and had a slight brownish colour to them.

What caused this?

Problem-solving process	The steps you need to solve the problem
1 **Identify the problem**	
2 **List your observations**	
3 **Make an inference or inferences**	
4 **Test your inference or inferences**	

Chapter 5
Our energy future

Problem solving

Problem solving strategy:

List all the possibilities and select the most likely.

Many scientific problems do not have one simple answer. In fact there may be several solutions to the problem. You need to list all the possible solutions, compare them using relevant criteria and then select the most likely.

Situation: Tom and Molly Dawson live on the outskirts of a large city. Their property is connected to electricity and gas. They have some firewood on their property, but not enough for a lifetime supply. They need to purchase a new stove and are concerned about the initial and ongoing costs, the reliability of electricity supply, and any long-term effects on the environment.

The longer Tom contemplated the wood pile, the better an electric stove looked.

Problem-solving process	Problem-solving example
1 **Identify the problem**	What type of stove should Tom and Molly buy?
2 **List all the possibilities**	• electric stove • gas stove • wood-burning stove
3 **Collect data to compare the advantages and disadvantages of the possibilities** Put this information into a table, chart or graph.	(see table below)

	Initial cost	Ongoing costs	Reliability of energy supply	Long-term environmental effects
Electric stove	\$700–800	\$15 average per 3 months	reliable unless electricity failure	coal supplies electricity and is non-renewable
Gas stove	\$700–800	\$20 average per 3 months	reliable	gas is a non-renewable resource
Wood-burning stove	\$3000–3500	\$10 average per 3 months	reliable, but limited long-term wood supply	wood is renewable but pollutes the air

4 **Select the most likely possibility to solve the problem. Justify your decision.**	All three stoves have advantages and disadvantages. The disadvantage of the wood-burning stove is its high initial cost and the fact that the wood supply, although cheap and a renewable resource, is not unlimited. The smoke and waste gases also pollute the atmosphere. The electric and gas stoves cost the same amount to buy. The electric stove has a fairly reliable energy supply, but power stations burn coal to produce electricity, and the waste gases are major pollutants of the atmosphere. The gas stove seems to be the best choice because it has a reliable energy source and the gas does not produce the same pollutants as burning wood or coal. These advantages outweigh the problems of gas having the highest ongoing costs and being a non-renewable resource.

You will use the same process to solve the problem at the end of this chapter.

During reading

Exercise 1 Writing clues

After reading pages 114–118

Read the paragraphs in the box and answer the questions below. Use Linking words on page 99 of this workbook for reference.

In contrast to fossil fuel power stations, nuclear power stations do not usually cause air pollution. The use of nuclear fuel not only conserves valuable fossil fuels, but also reduces greenhouse gas emissions and reduces the problem of acid rain. However, there are two major problems with the use of nuclear power: the disposal of highly radioactive wastes, and the risk of serious accidents. Both of these increase the costs of building and soperating nuclear power stations.

A nuclear power station uses a process called nuclear fission, that leads to approximately 25 tonnes of spent fuel being produced each year. Most of this spent fuel is uranium and plutonium. Both can be recycled, but there is almost 1 tonne of unused, highly radioactive material produced each year. These wastes can be solids, liquids or gases.

1 Select *Cause and effect linking words* from the words that are underlined ______

______

2 Select *Contrast linking words* from the words that are underlined ______

______

3 Select *Generalisation linking words* from the words that are underlined ______

______

4 What does *both* refer to in the first paragraph? ______

5 What does *both* refer to in the second paragraph? ______

Exercise 2 Writing clues

After reading pages 114–118

Write the clues for this crossword below using the information from the textbook.

				1 A	C	I	D	R	A	I	2 N			3 U		
											O			R		
					4 J						N			A		
5 T			6 G	L	O	B	A	L	W	A	R	M	I	N	G	
R					U						E			I		
A			7 P	O	L	L	U	T	I	O	N			U		
N					E						E			M		
S		8 C									W				9 S	
10 P	H	O	T	O	C	H	E	M	I	C	A	L	S	M	O	G
O		A									B				L	
R		L									L			11 G	A	S
T			12 F	O	S	S	I	L	F	U	E	L	S		R	

Across clues

1 ______________________________

______________________________ (2 words)

6 ______________________________

______________________________ (2 words)

7 ______________________________

______________________________ (2 words)

10 ______________________________

______________________________ (2 words)

11 ______________________________

____

12 ______________________________

______________________________ (2 words)

Down clues

2 ______________________________

____

3 ______________________________

____

4 ______________________________

____

5 ______________________________

____

8 ______________________________

____

9 ______________________________

____

Exercise 3 Translating from text to table

After reading pages 116–127

Use the information in the textbook to complete the table below and on the next page.

Energy source	Brief description of how it is harnessed	Advantages	Disadvantages
nuclear energy			
solar energy			
hydro-electric energy			

Energy source	Brief description of how it is harnessed	Advantages	Disadvantages
wind energy			
tidal and wave energy			
geothermal energy			
biomass energy			

Exercise 4 Persuasive speaking

After reading the whole chapter

In everyday life you often need to be persuasive. You might want to convince your parents to buy you a car. You might have to explain at a job interview why you are the best person for that job, or you might want to convince your friends to do something (or not to do something!).

In this exercise you will be learning about techniques used to make an effective persuasive speech.

1 **Researching a topic using several references**

Before you can write a persuasive speech, it is essential to do some research. Your audience will more likely be convinced if you can support your argument with facts and figures, and quote other sources that support your point of view. To research your topic properly you need to complete the following steps.

a Use the internet or select the books or other resources using the catalogue in the library.

b Locate the relevant information in the resources using the menu index, headings and sub-headings.

c Scan the information and decide how useful it is.

d If the information is relevant to your chosen topic, read it carefully and make your own notes.

e When you make notes, you should summarise the information in your own words. (As you compress the information into a few of your own words you understand the ideas better and develop your own language and thinking skills.) Do not copy other people's words straight from the source. This is not only plagiarism but also a mindless exercise.

2 **Judging and evaluating arguments**

Always check who runs the website or published the book and what the author's affiliations are. For example, the views of someone from the Australian Conservation Foundation about uranium will almost certainly be different from the views of someone from the Australian Mining Industry Council. Also, when you are reading and re-reading the information you have located, be aware of some of the techniques which biased authors may use to try to 'con' readers. Here are some examples of these techniques.

a Stating opinions as facts

The year 2020 will see the doom of civilisation …

b Name calling

Greenies are only dole-bludgers who …

c Over-generalising

All chemicals are bad for you …

d Over-use of emotive language

A white death is slowly and insidiously devouring millions of hectares of Australia's prime farming lands, rendering them sterile and lifeless …

e The peer pressure argument

Don't be left out — join our cause now and …

f Attacking the person instead of the argument

Bloggs says that uranium is potentially dangerous, but he's a fool anyway …

g The no-argument argument

Why shouldn't we chop down trees? Because it's wrong, that's why …

h Red herring (an irrelevant statement to take attention away from the issue)

Smog is a major problem in Los Angeles. A lot of film stars like Matt Damon work in Los Angeles. His films include …

3 **Preparing your persuasive speech**

The purpose of a persuasive speech is to convince your audience through carefully reasoned and logical argument that they should agree with your point of view.

a In the **introduction** of your speech:
- state your point of view clearly and firmly
- sum up the case you will present: this is called your contention

b In the **body** of your speech:
- develop the argument in clear steps
- support each main point with evidence
- take one side of the argument only—if you think it is necessary to mention a counter-argument, you should do so only if you can rebut it (present a strong case against it)

c In the **conclusion** of your speech:
- in strong, stirring language, state the logical conclusion which your audience must reach if they have agreed with the arguments in the body of your speech
- do not introduce any new ideas

4 **Presenting your speech**

As well as presenting a logical, factual, well-developed and well-researched argument, there are some techniques which you may like to consider to enhance your presentation in the box below.

Speech techniques

a **Emotive language**— Balance your logical approach with a sincere appeal to the personal and emotional side of your audience so that they feel involved (but don't overdo it as it will sound insincere).
e.g. Every one of us here today has a duty to protect our children's heritage.

b **Rhetorical questions** (questions to involve the audience but which the speaker answers)
e.g. What can we do? First of all, we …

c **Repetition**— e.g. Worldwide, over 100 species per day are now becoming extinct. These 100 species …

d **Cumulation** ('like' words grouped for effect) e.g. clean, healthy, invigorating air …

e **The 'three' effect**— e.g. I will, I must, I can do something …

f **Alliteration** (using words starting with the same letter) e.g. poisonous, putrid pollution

g **Shorter sentences**— because the listener cannot go back to check on what has been said (as the reader of an essay can).

The secret of a good speech is to prepare, prepare, prepare —and practise, practise, practise.

Your task

Choose one of the topics below to present as a persuasive speech. You should use information from your textbook and other reference material, as well as the information given previously in this exercise. Your audience will be your teacher and the class.

Your purpose is to persuade them through your logical arguments and vocal skills that your point of view is correct.

CHOOSE FROM THESE TOPICS

1. Burning petroleum is akin to firing up the kitchen stove with banknotes. (Mendeleev)
2. There's no alternative but the nuclear alternative.
3. You can't beat the sun for energy.
4. Don't be short-sighted—look for renewable energy sources.
5. Protecting the Earth doesn't mean going without. It means making better use of what we have.
6. Cars are environmental hazards.
7. We live on an island—so why aren't we using tides and waves to generate energy?
8. We are part of the Earth and it is part of us.
9. Fossil fuels or renewable resources? Is there really a choice?

Roundup

Exercise 5 Problem solving using the strategy:

List all the possibilities and select the most likely.

An energy supply system is to be built for an Antarctic tourist resort, but which energy source/s would be best? Your task is to analyse the data below and make a recommendation. Use the table on the opposite page to show your problem solving.

Tenders are called for the supply of an energy system for an

Antarctic Tourist Resort

Manufacturers and suppliers of energy systems are invited to register in an appropriate proposal their interest in tendering for the supply of energy to the proposed tourist resort at Mawson, Antarctica. Potential suppliers will be required to provide a general description of their proposed method of energy supply. The proposed method should be cost effective and suited to the severe weather conditions of Antarctica.

The resort will operate only during the summer months and will accommodate 50 tourists and staff initially, with growth to 100 over the next 5 years. The expected energy requirement of the resort is 100 kilowatts.

All proposals must include an environmental impact statement, and must respect the present moratorium on mining in Antarctica.

Companies interested in tendering are invited to send their general proposal to:

The Co-ordinator
Antarctic Adventures
GPO Box COLD
Sydney NSW 2001

Mawson physical and environmental data summary sheet

Typical weather for January	
Mean temperature	+1.0°C
Maximum temperature	+6.8°C
Minimum temperature	–4.3°C
Mean wind speed	28 km/h
Maximum wind gusts	109 km/h
Days of strong wind	21
Days of rain/snow	7
Average wave height at 65°S	3 m
Mean daily sunshine*	8 h

(*intensity only 40% of what it is in Sydney)

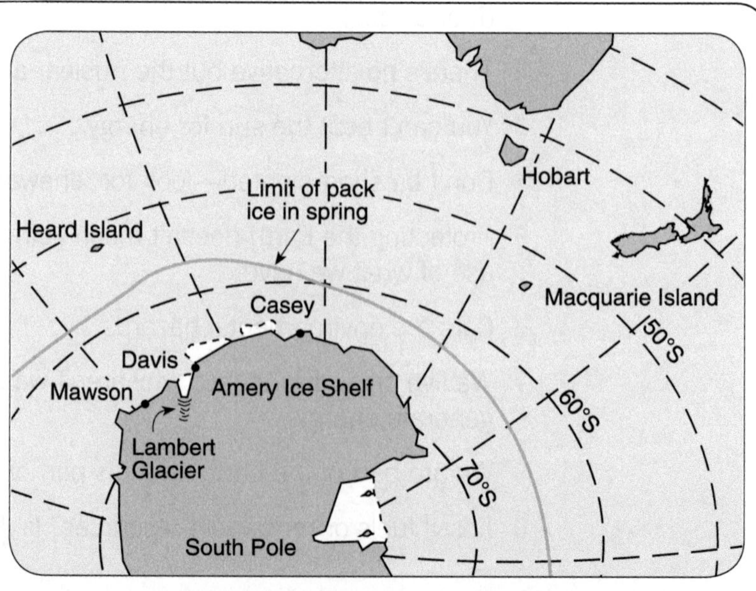

Problem-solving process	The steps you need to solve the problem
1 **Identify the problem**	
2 **List all the possibilities**	
3 **Collect data to compare the advantages and disadvantages of the possibilities**	
4 **Create a table listing the possibilities and the advantages/disadvantages of each**	
5 **Select the most likely possibility to solve the problem and justify your decision**	

Chapter 6
Space science

Problem solving

Problem solving strategy:

Use a rule or formula.

Sometimes you cannot solve a problem unless you know the correct scientific rule or mathematical formula.

Situation: Hannah and Charlie are rocket scientists.
They are developing a rocket engine to fire a small rocket into space.
Their rocket has a mass of 100 kg.

To begin their study, Hannah and Charlie need to calculate the smallest force the engine needs to develop to lift the rocket off the ground.

Problem-solving process	Problem-solving example
1 **Identify the problem**	What force must the rocket engine develop to lift the rocket off the ground?
2 **Identify the rule or formula needed to solve the problem**	In your previous studies you learnt that weight is a force. This force is dependent on the acceleration due to gravity. The acceleration due to gravity is 9.8 m/s^2. You also learnt in Chapter 2 that: force = mass x acceleration.
3 **Apply the rule**	The engine must develop a force that is greater than the weight of the rocket. The weight of the rocket is: weight = mass x acceleration = 100 kg x 9.8 m/s = 980 N So, to lift the rocket off the ground, the engine must develop a force greater than 980 N.

You will use the same process to solve the problem at the end of this chapter.

During reading

Exercise 1 Explaining a mathematical formula

After reading page 145

The acceleration of an object is proportional to the force acting on it, and inversely proportional to its mass. So, $a = \frac{F}{m}$.

The acceleration is directly ______ to the force. That's why the F is on top in the equation $a = \frac{F}{M}$. But, if you keep the force the same and double the mass then ______

And if you triple the mass, the acceleration ______

Exercise 2 Interpreting illustrations

After reading Rocket science on page 147 and doing the Experiment on page 148

Briefly describe the action and the reaction in each of the situations illustrated below.

Action: ______________________

____

Reaction: ______________________

____

Action: ______________________

____

Reaction: ______________________

____

Action: ______________________

____

Reaction: ______________________

____

Exercise 3 Rewriting text in your own words

After reading page 147

Newton's third law of motion is on page 147 of your textbook, and his first and second laws are on page 44 and 49. Use the table below to summarise these three laws in your own words and give an everyday example of each.

Newton's three laws of motion		
	What the law says	Everyday example
First law		
Second law		
Third law		

Exercise 4 Comparing and contrasting

After reading pages 149–150

1 Complete the table below.

	Solid-fuel engine	Liquid-fuel engine
Brief description		
Advantages		
Disadvantages		
Uses		

2 Use the table to complete these sentences describing similarities and differences between the two types of rocket engines.

a Both solid-fuel and liquid-fuel engines ____

____

b Solid-fuel and liquid-fuel rockets are different in that ____

3 Now join these sentences together to write a paragraph to compare and contrast solid-fuel and liquid-fuel rockets. Remember to start with a topic sentence.

____

____

____

____

____

____

____

____

Exercise 5 Recalling information

After reading Orbiting the Earth on pages 152–155 and doing Investigation 9

1 Define the following words, giving examples.

a An orbit is ______

______

______

b A satellite is ______

______

______

2 Write a generalisation linking the altitude of a satellite to its orbital speed. ______

______

3 Why does a polar orbiting satellite 'see' every part of the Earth's surface? ______

______

______

4 Explain in your own words why a geostationary satellite remains over the same point on the Earth's surface.

______

______

______

5 A merry-go-round has a scene painted on the central column which does not move. How would the merry-go-round need to be changed so that, as the children rotate, they see the same scene on the column?

______

______

______

______

______

______

______

Exercise 6 Reading critically

After reading Types of orbits on pages 154–155

1 Use the table to summarise the information in the text on the three types of satellites.

	Low Earth orbit	Polar orbiting	Geostationary orbit
Altitude			
Speed			
Advantages			
Disadvantages			
Uses			

Note: If there are any empty boxes you may need to do some library research.

2 Add these four labels to the diagram.

direction object will move if string breaks
downward weight of masses
force keeping object in orbit
orbit

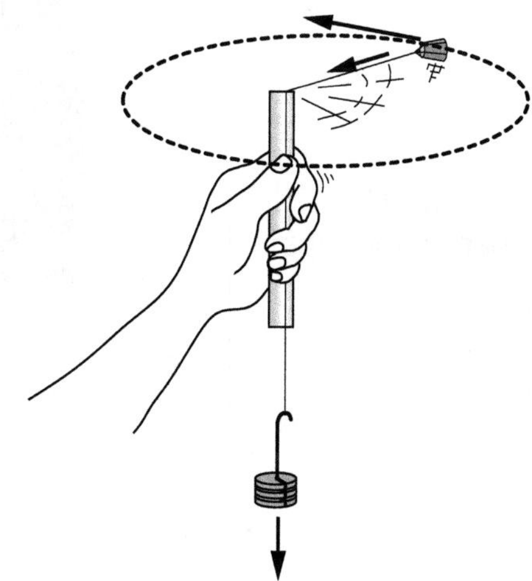

3 a Explain in your own words why an orbiting satellite does not fall back to Earth.

____

____

b What could cause an orbiting satellite to fall back to Earth? ______________________

____

____

Exercise 7 Writing in role

After reading Living in space on pages 158–162

It is the year 2019 and you are orbiting Earth in a Crew Exploration Vehicle at an altitude of 400 km.

Your task is to write a personal space email to a close friend for transmission to Earth. Your email should be about 300 words and describe your experiences.

1 To assist you in your task, complete the mind map below to write some ideas for inclusion in your email.

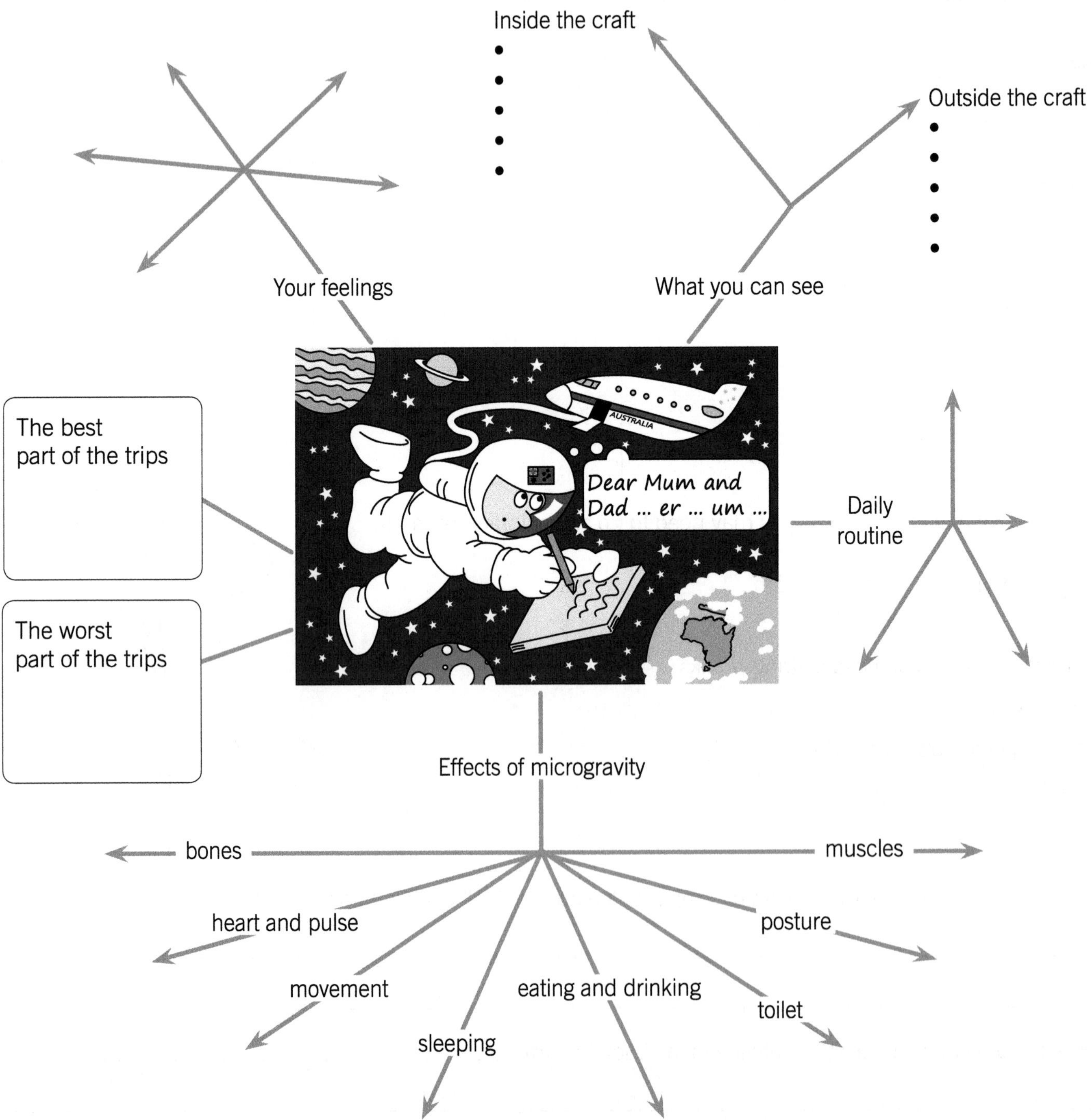

Roundup

Exercise 8 Problem solving using the strategy:

Use a rule or formula.

A new weather satellite is to be launched from the Cape York Space Port. It is to be placed in a geostationary orbit 35 786 km above the equator to the north of Papua New Guinea. (See page 155 of the textbook.)

When the satellite has been launched, what speed will be necessary to keep the satellite in geostationary orbit? Give your answer in m/s.

Problem-solving process	Problem-solving example
1 Identify the problem	
2 Identify the rule or formula needed to solve the problem	
3 Apply the rule	

Chapter 7
Periodic table

Problem solving

Problem solving strategy:

Use evidence to reach a logical conclusion.

When you need to make a decision or reach a conclusion, you need to investigate the facts and use evidence you gather to work logically through to the conclusion.

Situation: You are a manufacturer of boat sails. A famous sailor wants to build a new yacht for the upcoming Sydney to Hobart yacht race. He asks you to make a sail for his boat from a choice of eight different fibres. Which one will you choose?

Problem-solving process	Problem-solving example
1 **Identify the problem**	Use the data in the table on the following page to select the most suitable fibre for a sail.
2 **Study the data and make sure you understand it** In this example you need to know the meaning of each of the properties. You also need to know whether it is best to have a high number or a low number for each of the properties.	You have a choice of eight different fibres whose properties are listed in the table on the next page. You have to decide which properties you need, then make a first and second choice, and give reasons for your choice.

Properties of fibres	A	B	C	D	E	F	G	H
Density (g/cm^3)	1.31	1.55	1.34	1.52	1.30	1.14	1.38	1.16
Dry strength ($N/m^2 \times 10^4$)	1.1	3.2	3.6	2.4	1.2	4.5	4.1	2.4
Wet strength ($N/m^2 \times 10^4$)	0.9	3.4	2.6	1.3	0.8	3.8	4.1	2.0
% stretch before breaking	30–45	5–10	20–25	17–25	25–28	23–42	25–35	25–45
Force needed to stretch ($N/m^2 \times 10^4$)	29	53	63	64	36	20	60	40
% elastic recovery	100	70	90	60	85	100	80	90
Moisture absorbance (%)	16	9	11	13	3	5	0.5	2
Other properties	warm 'feel', may shrink a lot, moderate wear, can irritate skin	firm, comfortable, tough, may shrink a little, may crease	crisp attractive 'feel', very poor resistance to wear	soft, very good resistance to wear, crease-resistant	crease-resistant	soft, hard-wearing, crease-resistant, very tough	hard-wearing, very tough	soft, warm 'feel', non-shrink, tough, crease-resistant

3 Use the evidence to reach a logical conclusion

The properties that are important for making boat sails are as follows:

	Most suitable
Density needs to be low (light sails)	H and F
Dry strength needs to be high	F and G
Wet strength needs to be high since the sails are likely to get wet	G and F
Elastic recovery needs to be as close to 100% as possible so the sails keep their shape	F, A and H
Moisture absorbance needs to be as low as possible so the sails don't get too heavy	G and H

(To help in the final decision it is a good idea to pick the best two or three fibres for each property.)

You can now *use the information above to reach a logical conclusion.*

F would seem to be the most suitable fibre for making boat sails, because it is first or second best for four of the properties and its moisture absorbance is quite low.

G would be the second choice.

You will use the same process to solve the problem at the end of this chapter.

During reading

Exercise 1 Answering multiple-choice questions

After reading pages 170–174

In previous workbooks you have practised creating and answering multiple-choice questions. Remember a multiple-choice question has two parts:

- the stem which may be a question or an unfinished statement
- at least three possible answers, only one of which will be correct, with the others incorrect or only partly correct. Sometimes, one of the possible answers will be 'All of the above' or 'None of the above'.

Read the stem and the possible answers very carefully before making your decision. Then shade the letter which you think is the correct one to complete the sentence. The first one is done for you.

1 The periodic table used today was devised by—

(a) Albert Einstein ⟶ incorrect
(b) Andy Thomas ⟶ incorrect
(c) Dmitri Mendeleev ⟶ correct
(d) Dmitri Mendelayif ⟶ partly correct

2 The number of protons and the number of electrons in an atom—

(a) are always the same
(b) are always different
(c) always add up to 8
(d) none of the above

3 The atomic number of krypton is

(a) less than the atomic number of neon—
(b) the same as for all the other noble gases
(c) more than the atomic number of arsenic
(d) the same as the atomic number of xenon

4 Boron and tellurium are—

(a) both non-metals
(b) both metals
(c) both gases
(d) both semiconductors

5 The periodic table group number indicates—

(a) whether the element is a solid, liquid, gas or synthetic
(b) whether the element is a metal or a non-metal
(c) the total number of electrons in the atom
(d) the number of electrons in the outer shell of the atom

6 The fewer the electrons in the outer shell of an atom—

(a) the more reactive the element will be
(b) the less reactive the element will be
(c) the more stable the element will be
(d) the more likely the element is a metal

Exercise 2 Devising multiple-choice questions

After reading pages 170–174

Use the periodic table to devise your own multiple-choice questions. Use the same format and structure as in Exercise 1. *Make sure you have only one correct answer.*

Try your questions on members of your class.

1 ______________________________

(a) ____________________

(b) ____________________

(c) ____________________

(d) ____________________

2 ______________________________

(a) ____________________

(b) ____________________

(c) ____________________

(d) ____________________

3 ______________________________

(a) ____________________

(b) ____________________

(c) ____________________

(d) ____________________

Exercise 3 Using mathematical patterns

After reading pages 172–174

1 Find the pattern in the atomic numbers of the noble gases in Group 18.

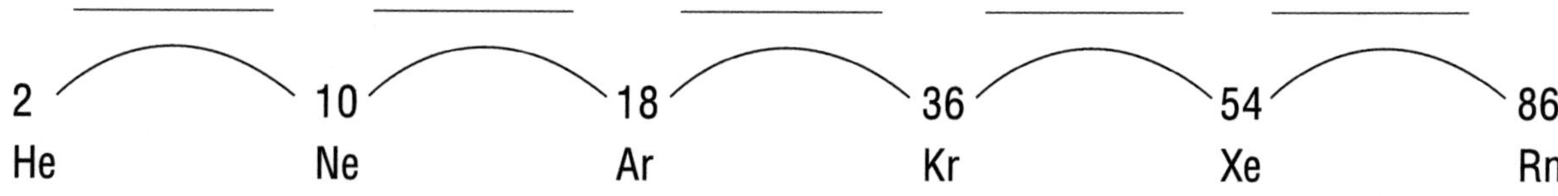

2 Imagine that another noble gas was found that fits directly under radon.

a Predict what its atomic number would be.____________________

b Name this imaginary element after yourself, someone else, your school, your suburb or town by adding the ending —on.____________________

Exercise 4 Identifying characteristics

After reading Metals on pages 176–179

Complete the overview by placing characteristics and examples under the relevant headings below.

Note: Some characteristics fit the general heading Metals while others apply specifically to the sub-headings Alloys, Reactive metals and Transition metals.

- mixture of 2 or more different metals
- packed in regular 3-D pattern called a lattice
- usually form coloured compounds when they react with non-metals
- very reactive, only found in nature as ionic compounds
- examples: iron, copper, zinc
- reactive but not as reactive as Group I metals
- very hard metals with a high melting point
- examples: brass, steel, 9 carat gold
- metallic shine or lustre
- found in the middle of the periodic table
- examples: magnesium, calcium
- good conductors of heat and electricity
- examples: sodium, potassium
- electrons can move freely in spaces between atoms
- Group I metals: a single electron in the outer shell
- Group II metals: two electrons in the outer shell
- malleable

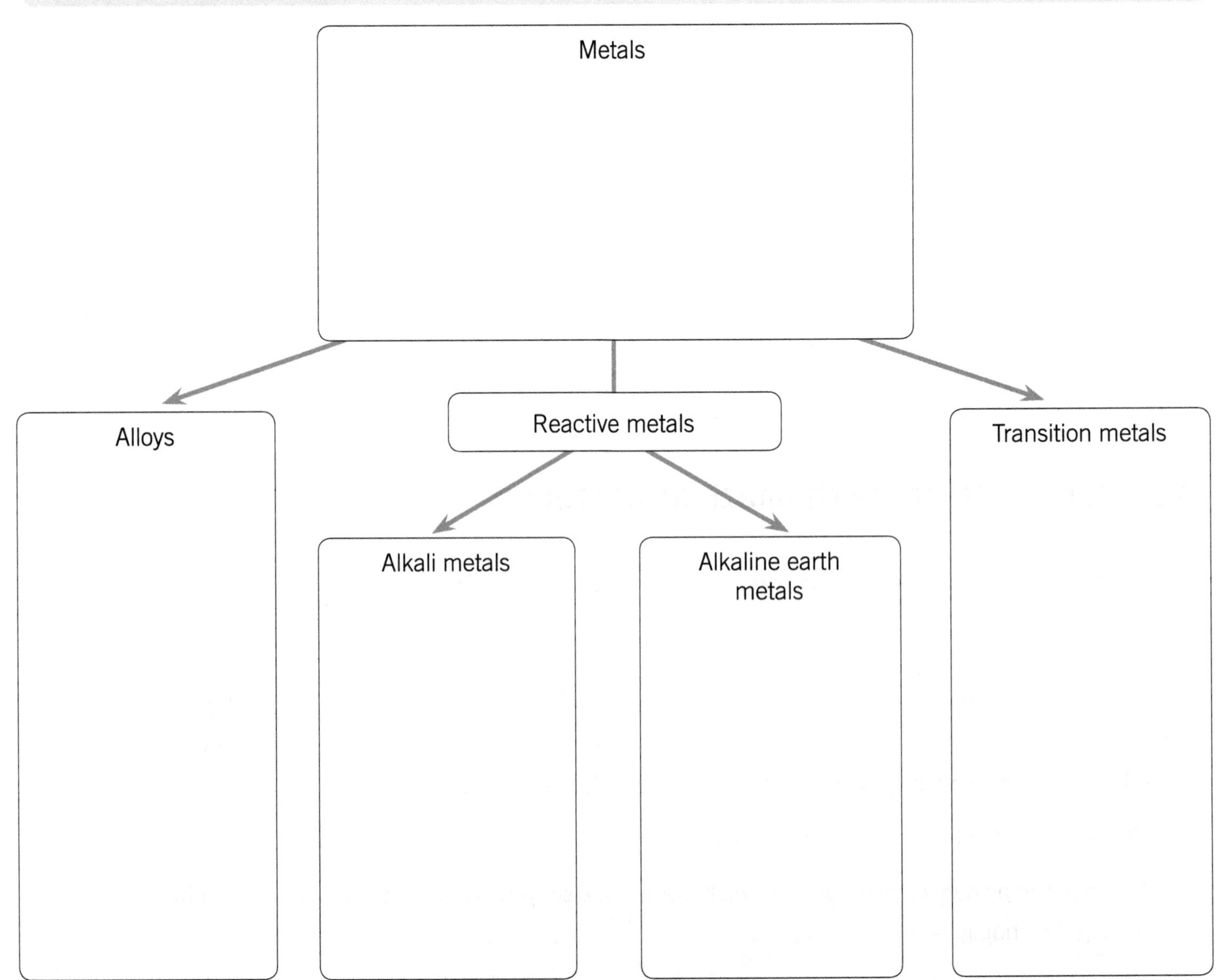

Exercise 5 Translating from text to other forms

After reading Non-metals on pages 180–182

1 Use the information in the textbook to complete the following table.

	Group 17 elements	Group 18 elements
Also called ...		
Degree of reactivity		
Three examples		
Other relevant information		

2 Use *Contrast linking words* from page 99 of this workbook to write a sentence contrasting the Group 17 and Group 18 elements.

____

____

____

____

3 Use a *Comparison linking word* and a *Contrast linking word* in one sentence to describe one similarity and one difference between graphite and diamond.

____

____

____

4 Complete the cartoon.

Exercise 6 Matching

After reading Extracting metals on pages 184–187

Rule lines to match the name of the extraction process with the description of the process and an example of a metal extracted using that process.

One set of matches is done for you as an example.

Name of extraction process	Description of process	Examples of metal extracted in the process
Froth flotation	The ore is converted to copper sulfide, which then reacts with oxygen to form sulfur dioxide and the metal.	Lead
Smelting	The metal is displaced from its ore in a solution by adding a more reactive metal. The more reactive metal and the metal to be extracted swap places.	Copper
Blast furnace	The metal ore is heated with carbon. The oxide is converted to the metal and the carbon combines with oxygen to form carbon dioxide.	Highly refined copper
Roasting	The metal ore is crushed and added to water, a chemical agent is added and air is bubbled through the mixture. The froth at the top containing the mineral is scraped off and the mineral dried.	Titanium
Electrolysis	The metal oxide, coke and limestone are fed into a furnace. Very hot air is blasted in the bottom, causing the coke to burn, forming carbon monoxide, which reacts with the metal oxide reducing it to the metal.	Copper and lead
Displacement	An impure metal plate connected to the positive terminal and a pure metal plate connected to the negative terminal are placed in a solution of copper sulfate and sulfuric acid. As the current flows, the impure metal dissolves and pure metal is deposited on the negative electrode.	Iron

Roundup

Exercise 7 Problem solving using the strategy:

Use evidence to reach a logical conclusion.

PART A
Geologists suspect there is a deposit of copper ore in the Currawong Creek area. Water samples from four points on the map were analysed. The samples from A and B showed traces of copper, but C and D did not.

Where do you think the ore deposit is likely to be? Mark this on the map on the right. Explain your answer.

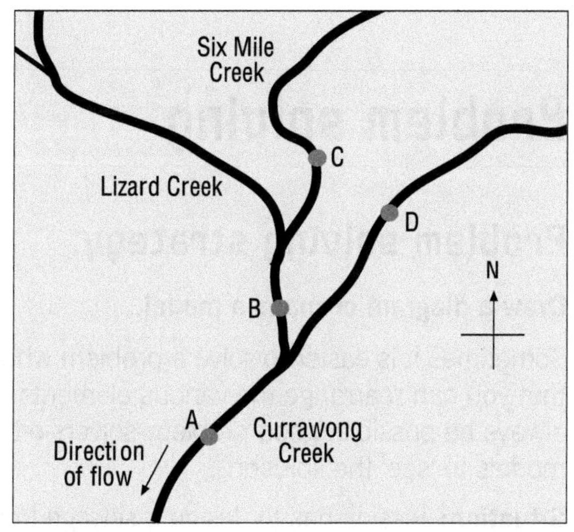

PART B
The search was narrowed to the Lizard Creek area. Soil samples were taken at various points in the area thought most likely to contain ore. The table on the right gives the map references of locations where medium, high and very high concentrations of copper were found.

Use the figures in the table to plot on the grid below the positions where medium, high and very high concentrations of copper were found. Use different colours for the different concentrations.

Copper concentration	Locations
medium	C3 C5 D3 D6 E3 E6 F7 G4 G7 H5 H7 I6 I7 I8 I9 J7 K8 K10 L9 L11
high	C4 D5 F4 F6 G5 G6 H6 J8 J9 K9 L10 M10
very high	D4 E4 E5 F5

PART C
The next step in the exploration is to drill to collect samples to find out the extent of the deposit. Your budget allows only three holes. Give the map references of the three places you would drill. Justify your decision.

Problem-solving process	Your solution
1 **Identify the problem**	
2 **Study the data and make sure you understand it**	
3 **Use the evidence to reach a logical conclusion**	

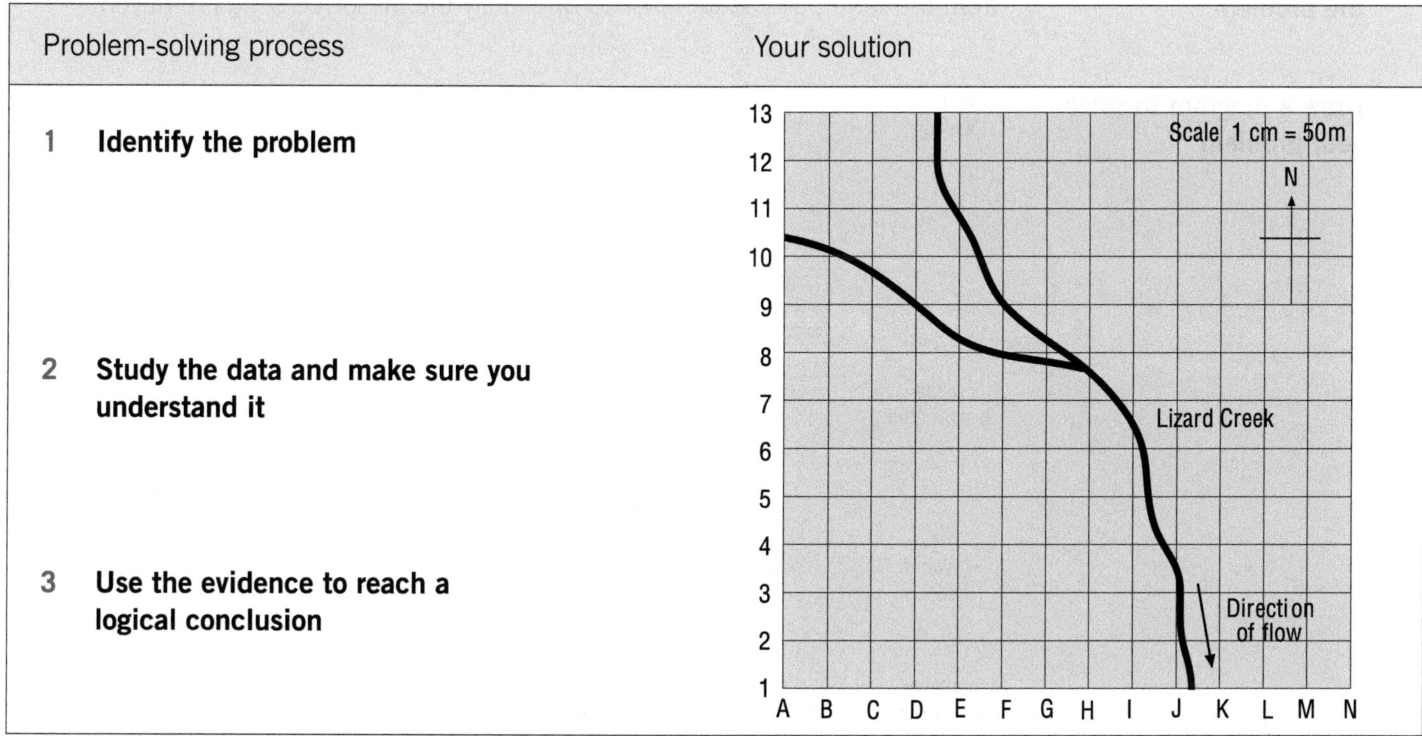

Chapter 8

Earth systems

Problem solving

Problem solving strategy:

Draw a diagram or make a model.

Sometimes it is easier to solve a problem when you can actually 'see' it so that you can rearrange the various elements or parts. While this may not always be possible, good problem-solvers often draw diagrams or make models to 'see' the solution.

Situation: Jessica has to design a sewage treatment system which will produce water which is acceptably clean. In the process the solids in the sewage have to be collected.

Problem-solving process	Problem-solving example
1 **Identify the problem**	What is needed in the sewage treatment process to produce water that has a relatively low bacterial count.
2 **Identify what you know that might help you solve the problem**	The raw sewage is mainly water that contains dissolved pollutants and suspended solids and bacteria. The sewage needs to have the solids removed from the water, the bacteria killed and lastly the dissolved solids removed.
3 **Draw a diagram to solve your problem**	raw sewage settling tank liquid containing small solids large solids settle solids collected filtration solids collected chlorine added acceptably clean water bacteria killed by chlorine

Exercise 1 Compiling a graphic outline

Knowing how a chapter is organised helps you to understand the relationships among ideas, and it is easier for you to find relevant information quickly.

In this chapter you will be asked questions that require you to locate information in the textbook. You can work through the chapter at your own pace.

To start with, complete the table below.

Main Headings	Sub-headings (List each sub-heading)	Figures (List the figures that have a caption)	Activities and investigations (List the title or describe what it's about, and give the page number)
8.1 Four spheres	The cycling of chemicals	Fig 2	Activity page 199: The carbon cycle
8.2			
8.3			

During reading

In Chapter 7 of this workbook, you learnt how to answer and write multiple-choice questions.

This is a very common method of testing your understanding in science, so the following exercises will give you lots of practice!

Stop at the end of each exercise to check your answers on page 68 of this workbook. Do not continue until you know exactly why any of your answers were incorrect by re-reading the textbook or checking with your teacher.

Use a pencil to shade the circled letter next to the answer you think correctly completes the statement.

For example— Insects have
(a) four legs
(b) six legs
(c) eight legs

Exercise 2 Answering multiple-choice questions

After reading Four spheres on pages 197–201

1 The lithosphere includes—
(a) the Earth's crust, mantle and core
(b) the rocky part of the Earth
(c) the Earth's crust and upper layers of the mantle

2 The hydrosphere—
(a) includes all the water on the Earth and in the atmosphere
(b) includes all the water on the surface of the Earth as well as groundwater
(c) both (a) and (b)

3 As more and more carbon dioxide dissolves in the oceans, the acidity of the water—
(a) increases
(b) decreases
(c) stays the same

4 Carbon is removed from the atmosphere when—
(a) dead organisms decompose
(b) plants and plankton photosynthesise
(c) fossil fuels are burnt

5 Carbon is recycled into the atmosphere when—
(a) coal, oil and wood are burnt
(b) decomposers break down the bodies of dead organisms into carbon dioxide
(c) both (a) and (b)

6 Nitrogen from plants and animals is recycled into the air through—
(a) the breakdown of protein in animal cells
(b) the action of bacteria in the soil
(c) lightning strikes

7 Nitrogen *fixing* means—

(a) nitrogen is made into protein by animals and plants

(b) nitrogen is converted to proteins by bacteria and some algae

(c) nitrogen is converted to nitrates and ammonium ions by bacteria and algae

8 Lightning can—

(a) convert nitrogen atoms and oxygen atoms to nitrogen oxides

(b) convert carbon atoms and oxygen atoms to carbon dioxide

(c) convert nitrogen atoms and oxygen atoms to ammonia

Check your answers on page 68 of this workbook.

Exercise 3 Answering multiple-choice questions

After reading Effects of human activity on pages 202–207

1 Biodegradable wastes—

(a) include animal wastes and decaying plants and animals

(b) can never be broken down

(c) include factory wastes such as sewage and heavy metals

2 Eutrophication in a lake is occurring when—

(a) there is a huge growth of decomposing bacteria

(b) there is a rapid decrease in the dissolved oxygen

(c) both (a) and (b)

3 The amount of dissolved oxygen—

(a) is not affected by water temperature

(b) is greater at 40°C than at 20°C

(c) is greater in cold water than in warm water

4 Examples of non-biodegradable wastes include—

(a) wastes from food processing factories

(b) wastes from steel processing factories

(c) wastes from meat processing factories

5 The concentration of non-biodegradable wastes—

(a) increases as you go up a food chain

(b) is higher in sharks than in a bait fish

(c) both (a) and (b)

Check your answers on page 68 of this workbook.

Exercise 4 Answering multiple-choice questions

After reading The atmosphere on pages 208–213

1 The atmosphere surrounding Earth reaches a height of approximately—

(a) 1000 m

(b) 600 m

(c) 1000 km

2 Clouds are formed in the—

(a) troposphere
(b) stratosphere
(c) ionosphere

3 Heat is trapped in a greenhouse because—

(a) short-wavelength solar radiation can pass through the glass, but longer-wavelength infra-red radiation cannot
(b) short-wavelength solar radiation cannot pass through the glass, but longer-wavelength infra-red radiation can
(c) the plants trap the sun's heat

4 Greenhouse gases are gases which—

(a) stop infra-red radiation getting through to Earth
(b) stop short-wavelength solar radiation getting through to Earth
(c) absorb infra-red radiation on its way back to space and re-radiate the heat in all directions, including some back to Earth

5 Burning fossil fuels and forests produces the greenhouse gas(es)—

(a) carbon dioxide
(b) nitrous oxide
(c) both carbon dioxide and nitrous oxide

6 The higher the concentration of greenhouse gases in the atmosphere—

(a) the higher the global surface air temperature
(b) the lower the global surface air temperature
(c) there is no effect on the global surface air temperature

7 The chemical symbol for ozone is—

(a) O
(b) O_2
(c) O_3

8 Ozone absorbs UV radiation, so the less ozone there is in the atmosphere—

(a) the greater the risk of skin cancer, diseases of the immune system, and eye problems
(b) the greater amount of UV radiation reaches the Earth
(c) both (a) and (b)

Answers

Exercise 2
1 c
2 b
3 a
4 b
5 c
6 b
7 c
8 a

Exercise 3
1 a
2 c
3 c
4 b
5 c

Exercise 4
1 c
2 a
3 a
4 c
5 c
6 a
7 c
8 c

Exercise 5 Matching words with meanings

After reading pages 202–204

Match each word with its meaning.

Words	Meanings
sewage •	• wastes that do not break down or which break down very slowly
biodegradable wastes •	• a reduction in the oxygen concentration in water, caused by decomposing organisms following excess plant growth
biological oxygen demand •	• water and wastes that come from kitchen sinks, bathrooms, laundries, toilets and industries
eutrophication •	• the build-up in concentration of a chemical as it moves from organism to organism in a food chain
thermal pollution •	• a measure of the quantity of dissolved oxygen required by the bacteria to oxidise organic wastes
non-biodegradable wastes •	• wastes that can be broken down by the micro-organisms in water
biomagnification •	• a soil condition in which salts dissolved in groundwater are brought to the surface through removal of trees or because of irrigation
salinity •	• a rise in the temperature of water affecting the ability of fish to reproduce and survive

Exercise 6

After reading pages 202–204

You are visiting your uncle Duncan who owns a farm. You notice that he seems to be using a lot of pesticides and fungicides, particularly one fungicide that contains traces of the heavy metal, mercury. Use the information in your textbook, especially the process of biomagnification, to explain to him why you are concerned that his excessive use of these chemicals could lead to the extinction of certain species of large fish in the ocean off the coast of Australia.

Exercise 7 Recalling words and meanings

In this chapter, there were many scientific words for you to learn and understand. Complete the table by adding the missing letters in the right-hand column or writing the meanings in the left-hand column.

Meaning	Word or words
1 the part of the Earth that includes the crust and upper part of the mantle	l _ _ h _ s _ h _ _ e
2 aquatic micro-organisms that produce oxygen in photosynthesis	_ _ y t _ _ l _ _ k _ _ _
3 substances containing nitrogen, found in organisms	_ _ o _ _ i _ s
4 a water pollutant that contains human wastes, food scraps, grease and dirt	_ _ _ a _ e
5	biodegradable
6	eutrophication
7	groundwater
8 the atmospheric layer that contains charged particles	_ o _ o _ _ _ e _ _
9 radiation with a longer wavelength than visible light	i _ _ _ a-_ _ d
10 general name for gases that trap heat in the atmosphere	_ r _ _ _ h _ _ _ _
11 the process that describes the heating of the Earth due to increased levels of greenhouse gases	_ _ o _ _ _ w_ _ _ i _ _
12	ozone
13	CFCs

Exercise 8 Writing a problem–solution essay

Here is a model of how to structure a problem–solution essay. Read it carefully and note how it is organised.

Title (usually a summary of the author's solution)	GREENHOUSE GASES MUST BE REDUCED
The problem	Greenhouse gases are responsible for trapping heat and sending it back to the Earth's surface. This increase in heat is gradually warming our globe which, in the not too distant future, will have dramatic effects on our climate and sea levels.
Supporting data or evidence for the problem	Over the last 100 years, the global surface air temperature has increased 0.7°C, resulting in a rise in sea level of 11–12 cm. Between the years 1800 and 2007, carbon dioxide, methane, and nitrous oxide emissions have increased markedly.
Author's position	This situation cannot be allowed to continue.
Evidence for author's position	If we continue to release greenhouse gases at the current rate, the Earth's temperature could increase by 2–4°C by 2030 and sea levels could rise between 20 cm and 1.4 m. Natural disasters such as floods, droughts and cyclones could occur more frequently.
Author's solution	A world-wide solution is the only answer. All countries must agree to reduce greenhouse gas emissions. Countries that cannot afford to do so must be assisted by more wealthy countries.
Author's evaluation	Only a global response now can avoid a disaster for future generations.

Now it's your turn to write a problem–solution essay. In your notebook, use the same structure to write an essay about the hole in the ozone layer.

Roundup

Exercise 9 Problem solving using the strategy:

Draw a diagram or make a model.

Situation: The curator of the Sydney Royal Botanic Gardens wants to grow some rare tropical plants that need a warm, constant temperature all year round. The plants also need full sunlight, so cannot be grown in the shade.

Suppose you were asked to design a greenhouse for the curator. The building has to be energy efficient and cannot be air conditioned as this dries the air too much for the plants.

Use what you have learnt in this chapter to draw diagrams to solve the curator's problem. The curator knows nothing about types of solar radiation, so make sure you write notes to explain how the greenhouse will work.

Problem-solving process	Problem-solving example
1 **Identify the problem**	
2 **Identify what you know that might help you solve the problem**	
3 **Draw a diagram to solve your problem.** (You might have to use your notebook as well)	

Chapter 9

Evolution

Problem solving

Problem solving strategy:

View the problem from different perspectives.

In some cases, the solution to a problem is not clear because it depends on how each person involved views the issues. Each person involved will see the best solution as the one which agrees with their own perspective. For this reason, when trying to solve problems that are controversial, you need to examine as many points of view as possible and try to work out the best solution.

Situation: Designer babies? By breeding and selecting animals and plants that have desirable features, breeders have been able to produce organisms that are highly valued. We could also selectively breed babies whose features are highly prized, e.g. beauty, intelligence and sporting ability. What are the arguments for and against this?

Problem-solving process	Problem-solving example
1 **Identify the problem**	Should we be able to selectively breed babies whose features are highly prized, e.g. beauty, intelligence and sporting ability?
2 **Identify people involved in the problem who would view the issues differently** Form a group with each member taking a different role and arguing the issues from their perspective OR as an individual. Think about the issues from each perspective. (This is called the subjective approach.) The example on the right shows six different speakers' perspectives.	Speaker 1: Children with beauty and intelligence will certainly have a better life. Speaker 2: Whatever happened to falling in love and reproducing your own strengths and weaknesses? Speaker 3: Babies aren't a scientific product, they are a gift from God and nature. Speaker 4: Children need to be nurtured by real parents, not people who want designer babies. Speaker 5: It's up to the individual to decide, not anyone else. Speaker 6: If it's scientifically possible, why not?

Problem-solving process	Problem-solving example	
3 List the arguments As each perspective is presented, list the arguments for and against designer babies.	**Speakers for** 1 Lifestyle of child 5 Individual's right to choose 6 Why not if scientifically possible?	**Speakers against** 2 The natural approach 3 The religious approach 4 Real parents
4 Examine the arguments objectively Try to be as removed from the subjective arguments and examine the problem and arguments presented on the facts or data only. (This is called the objective approach.)		
5 Make a decision In groups using consensus decision-making, or as an individual, make a decision which takes into account all perspectives.	Discuss the problem and make your decision by consensus. The guidelines for consensus decision making are: • All members should agree with the final decision. • Be courteous and reasonable in your discussion. • Make sure all group members contribute. • Listen to the opinions of others. • Remember that differences of opinion are useful in clarifying issues. • Avoid making a decision by majority vote or averaging. • Do not change your mind simply to avoid conflict or friendship tensions.	

You will use the same process to solve the problem at the end of this chapter.

During reading

In the next six exercises, you will find two sections:

- a Core exercise to test your knowledge and understanding
- a Challenge exercise to use and apply your knowledge and understanding

After reading each section in the textbook, complete the Core exercise and check your answers on page 81 of this workbook. If your answers are all correct, go on to the Challenge exercise. If you are unable to do the Core exercise, or your answers are incorrect, read the textbook section again to find out why you were wrong (and ask for help from your teacher if you need to). After this, go on to the Challenge exercise.

Exercise 1

After reading pages 223–225

Core exercise: completing an overview

Complete the overview on the next page using words from the box below.

recombination	independent assortment
weather and soil conditions	mutations
environmental factors	genetic factors
nutrition and health	(spontaneous formation of new and altered genes)

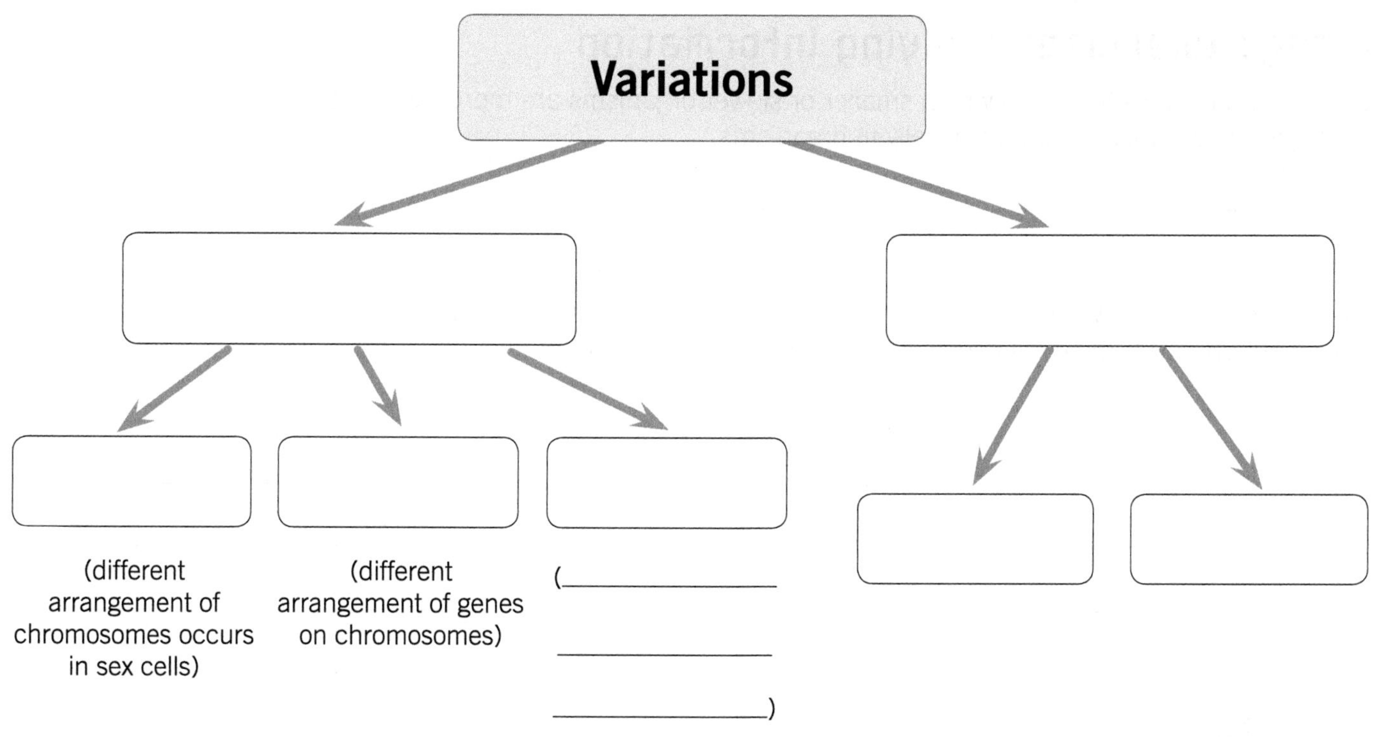

Challenge exercise: explaining

1 In your own words, explain the difference between the potential phenotype of an organism and the actual phenotype.

2 Which *genetic* factor(s) may explain how an elephant can be born smaller and weaker than its potential phenotype? ____

3 Which *environmental* factor(s) may explain how an elephant can be born smaller and weaker than its potential phenotype? ____

Exercise 2

After reading page 227

Core exercise: making and justifying decisions

Place a ✔ next to the statements that you think are correct, and a ✘ next to those that are incorrect. You should be able to justify your decision.

1 Organisms produce more offspring than their habitat can support, so only the best adapted will survive.

2 Natural selection means that individuals that are the physically fittest will survive.

3 Availability of water is a biotic selection agent.

4 Organisms that are adapted to their environment are likely to have offspring that will also have these favourable characteristics.

5 The competitive behaviour of male antelopes is an example of an abiotic selection agent in the environment.

Challenge exercise: applying information

1 Give an example of a situation in which smaller or slower organisms are more likely to survive than larger, more muscular and physically fit organisms.

____

____

2 What characteristics do weeds have that enable them to survive the selection agents in the environment and even the most aggressive gardener?

____

____

____

____

____

Exercise 3

After reading pages 232–233

Core exercise: matching terms and meanings

Match the following terms with their meanings.

1 species	a spontaneous alterations to the genes or the number of chromosomes in a cell
2 gene pool	b a process in which species change over time and develop into new species
3 mutation	c a population of organisms that have similar features and can interbreed
4 evolution	d the sum of all the genes in the population of a particular organism

Challenge exercise: hypothesising

In your textbook, you were given a hypothetical situation in which a population of land snails separated, became isolated, and then formed two different species over a long period of time.

Create your own hypothetical situation to explain the changes that occurred to the fish in two very large lakes over many thousands of years, as shown in the map below. Write your answer in your notebook.

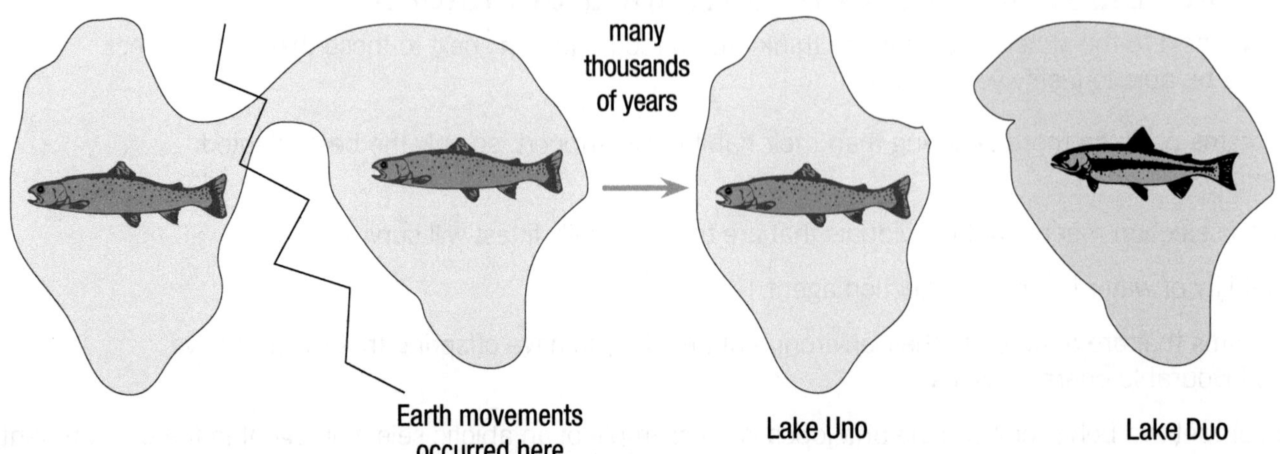

Exercise 4

After reading page 234

Core exercise: drawing time lines

1 A time line is a visual representation of a sequence of events. A time line is drawn horizontally or vertically and has marked intervals to indicate the units of time.

a What is the time interval on the following time lines?

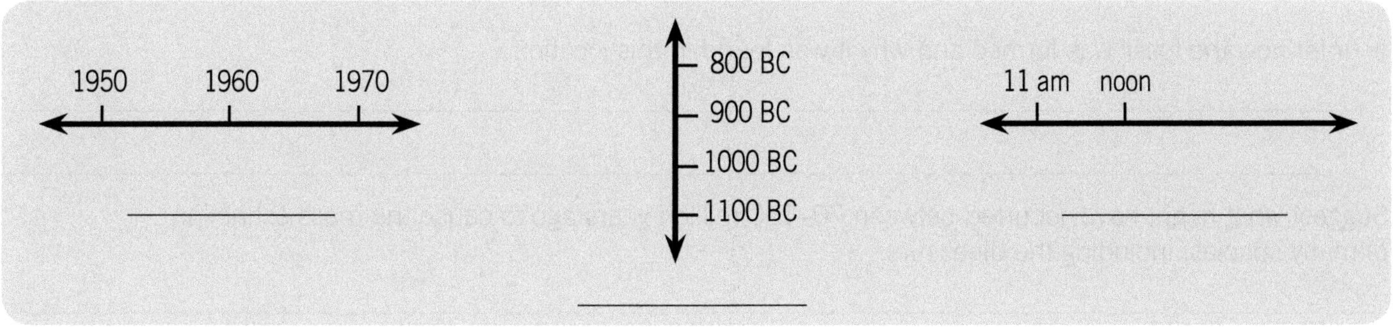

b Why are there arrows at each end of the time lines? ____________

____

c On the time line below, mark AD 152.

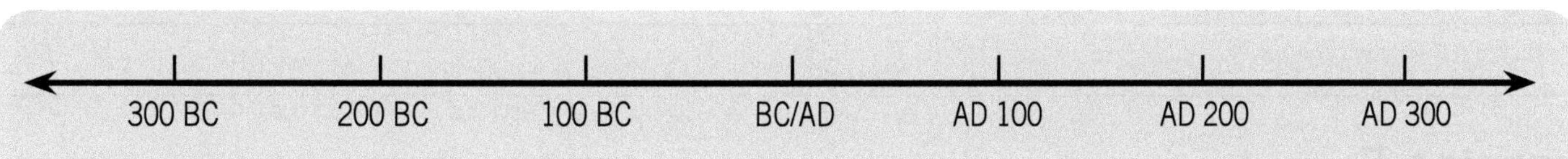

2 The vertical time line below contains information from the table on page 234 of your textbook. Each interval is 10 million years. Write in the box which fossils have been found in rocks formed at that time.

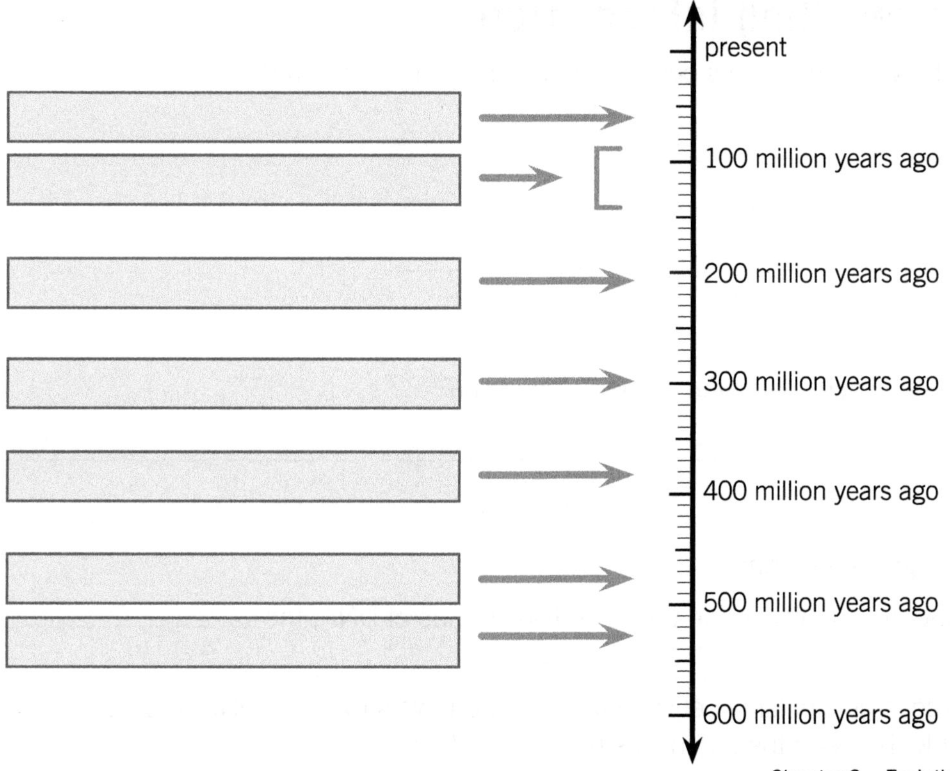

Challenge exercise: inferring

1 A geologist finds a fossilised marine shell in the form of a cast in the middle of an Australian desert. The sedimentary rocks in which it is found are dated at 400–450 million years old.

a Does the age of the fossil 'fit' the table on page 234 of your textbook? Explain your answer.

______

______

b Infer how the fossil was formed and why it was found in this location.

______

______

2 Suggest what might have occurred between 70–150 million years ago to cause the mass extinction of many species, including the dinosaurs.

______

______

______

______

______

______

Exercise 5

After reading pages 234–237

Core exercise: recalling information

1 List four ways in which evolutionary links between organisms can be inferred.

a ______

b ______

c ______

d ______

2 Which four land masses do scientists infer formed Gondwana?

______

______

3 Complete the following generalisations.

a The greater the difference between the bases on two strands of DNA, the ______ tightly the strands will bind.

b The closer the percentage similarity between the bases on two strands of DNA, the ______ the evolutionary links between the organisms from which they came.

Challenge exercise: evaluating

In 1859 Charles Darwin suggested that species change over time and develop into new species through the process of natural selection.

What strand of biology was not available to Darwin when he suggested this hypothesis? How does it help to support his hypothesis? ____

____

____

____

Exercise 6

After reading pages 239–245

Core exercise: matching examples

Match the examples on the right with the name of one of the three processes on the left.

1 Artificial selection

2 Biotechnology

3 Cloning

a A cell is taken from a drought-resistant cow and placed into an ovum from a high milk-producing cow.

b The gene from a disease-resistant red pawpaw is inserted into the chromosome of a yellow pawpaw which is more popular with consumers.

c A yellow thornless rose is crossed with a white thorned rose to produce a cream-coloured thornless rose.

d A champion stallion is mated with a thoroughbred mare.

e The genes from ice-forming bacteria that live in oranges are removed to prevent the oranges freezing in very cold weather.

f A gene that produces an insecticide in a particular weed is placed into the nucleus of a tobacco plant cell.

g A brown feathered hen is mated with a white rooster.

h A cell from an Olympic athlete is placed in the ovum of a nuclear physicist.

Challenge exercise: translating from text to table

After reading pages 239–245

Use the information in the textbook to complete the table below.

	Artificial selection	Genetic engineering	Cloning
Another name for this process			
Description of this process			
Advantages of this process			
Disadvantages of this process			
An example of this process			

Answers

Exercise 1

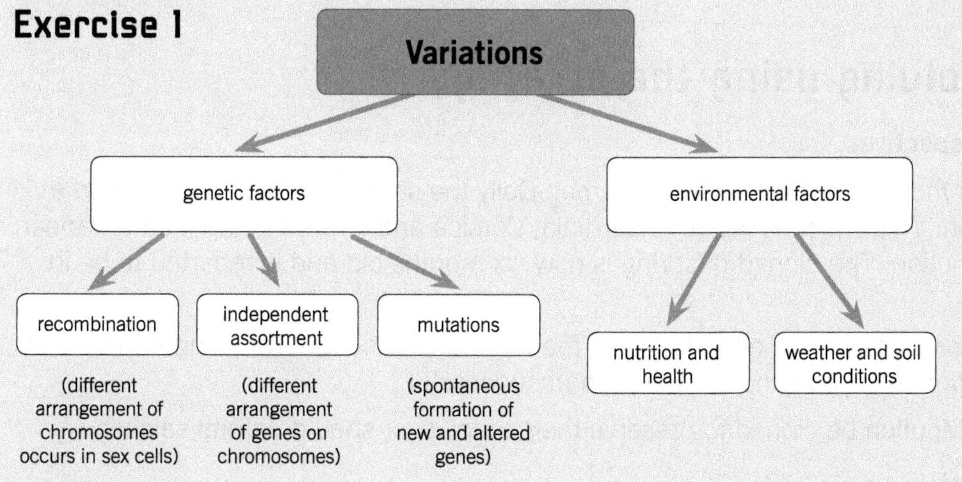

Exercise 2

1 ✔
2 ✗
3 ✗
4 ✔
5 ✗

Exercise 3

1 c
2 d
3 a
4 b

Exercise 4

1 a 10 years
100 years
1 hour

b The arrows show that the time line extends in both directions.

c

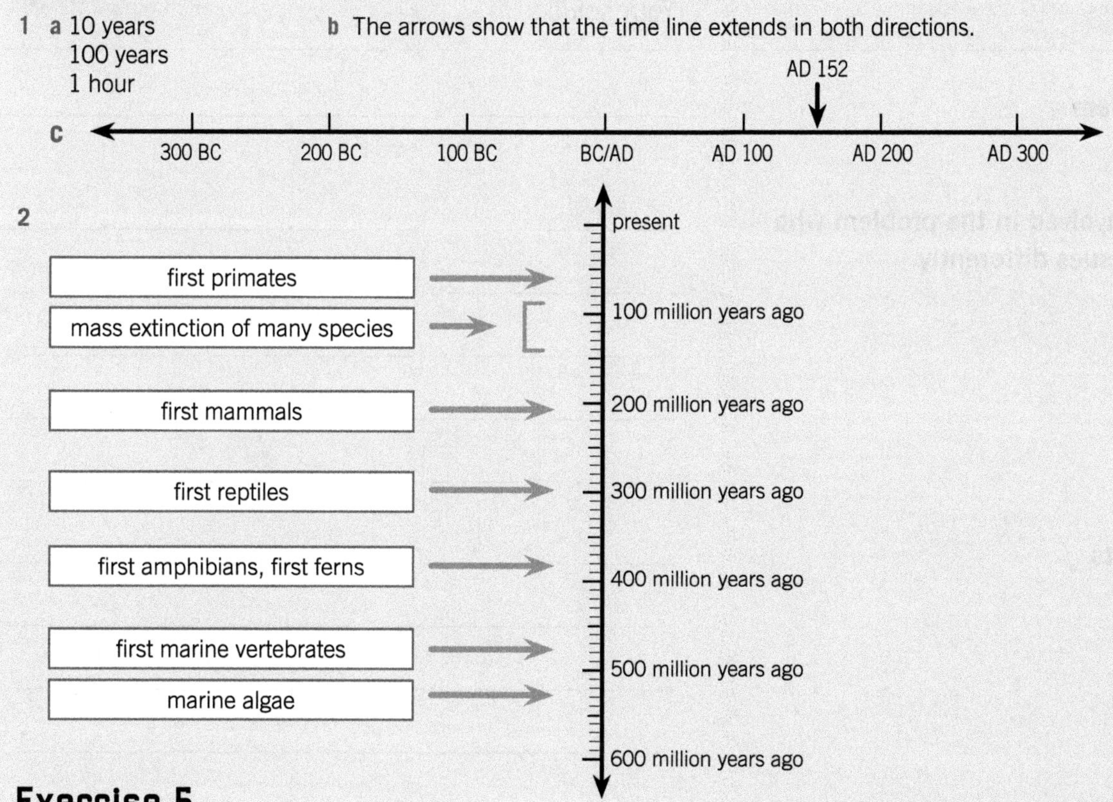

2

Exercise 5

1 a by studying fossils and estimating when the organism lived on earth
b by studying the distribution of different types of living organisms and fossils (biogeography)
c by comparing the embryos of different types of vertebrates
d by comparing the DNA of similar living species with that of fossils

2 Australia, Antarctica, South America and Africa formed Gondwana.

3 a The greater the difference between the bases on two strands of DNA, the less tightly the strands will bind.
b The closer the percentage similarity between the bases on two strands of DNA, the closer the evolutionary links between the organisms from which they came.

Exercise 6

1 artificial selection **c d g** 2 genetic engineering **b e f** 3 cloning **a h**

Roundup

Exercise 7 Problem solving using the strategy:

View the problem from different perspectives.

Situation: The scientists who created the world's first cloned mammal, Dolly the sheep, have recently cloned a wild species of sheep called a Mouflon. This sheep is native to Sardinia, Corsica and Cyprus in the Mediterranean Sea, and is in extreme danger of extinction. The cloned offspring is now six months old and is reported to be in good health.

Mouflon sheep have been kept in a zoo in Italy in an effort to breed them and increase the dwindling numbers in the wild. However, breeding the Mouflon in captivity has so far been unsuccessful.

Should endangered animals like the Mouflon be cloned to preserve their species, or should natural selection occur as it has with other wild species?

Problem-solving process	Your solution
1 **Identify the problem**	
2 **Identify people involved in the problem who would view the issues differently**	
3 **List the arguments**	
4 **Examine the arguments objectively**	
5 **Make decisions**	

Chapter 10
Exploring the universe

Problem solving

Problem solving strategy:

Select relevant information.

When you are asked to solve a problem relating to a particular topic, there is often so much information about the topic that it is easy to become confused. Before researching a topic to find the information, list what you are actually looking for. This is the relevant information, and nothing else is needed.

Situation: Imagine a spaceship can travel at twice the speed of light. How long would it take to travel from the star in the Southern Cross closest to the Earth to the star in the Southern Cross most distant from Earth?

Problem-solving process	Problem-solving example
1 **Identify the problem**	How long would it take a spaceship to travel from the star in the Southern Cross closest to the Earth to the star in the Southern Cross most distant from Earth?
2 **Identify the information that you need to solve the problem (the relevant information)** Ignore all other information. Do not allow a large amount of information to confuse you.	You need to know • the star in the Southern Cross closest to the Earth; • the star in the Southern Cross most distant from the Earth; • the distance between these two stars; • the formula for calculating time when you know distance and speed. You need to assume that • the Earth and the two stars lie on a straight line.

Problem-solving process	Problem-solving example
3 Find the information you need	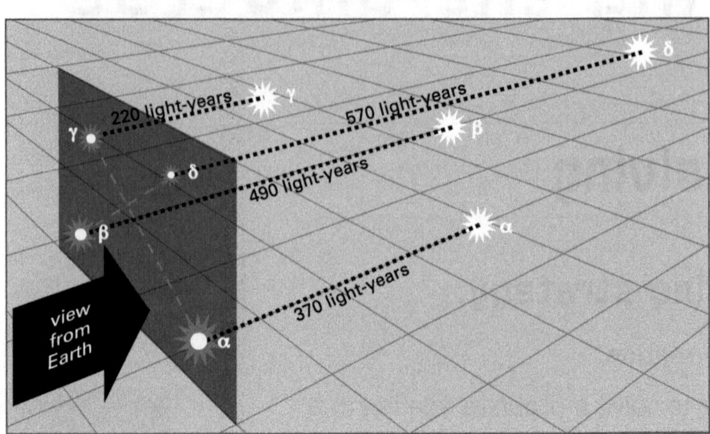
4 Using only the relevant information, work out the solution to your problem	• The star in the Southern Cross closest to Earth is 220 light-years from Earth. • The star in the Southern Cross most distant from Earth is 570 light-years from Earth. • The distance travelled between these two stars is 570 – 220 = 350 light-years • $\text{time} = \frac{\text{distance}}{\text{speed}} = \frac{350 \text{ light-years}}{2} = 175 \text{ years}$

During reading

Exercise 1 Calculating

After reading pages 252–253

When you need to write out a big number like the number of stars in our galaxy, it is very time-consuming to write it out in full. Instead you can use a special shorthand, *using powers of ten.*

$10 = 10^1$

$100 = 10 \times 10 = 10^2$

$1000 = 10 \times 10 \times 10 = 10^3$

$10\,000 = 10 \times 10 \times 10 \times 10 = 10^4$ and so on

Can you see the pattern? The small number above the 10 tells you the number of zeros there are. For example, the number of stars in our galaxy (100 000 000 000) can be written as 10^{11}.

How can you use this shorthand for the distance light travels in one year? First you write it in two parts:

$9\,500\,000\,000\,000 = 9.5 \times 1\,000\,000\,000\,000$

9.5: This part is always between 1 and 10.

1 000 000 000 000: This is what you have to multiply by to get your number.

You then write this as 9.5×10^{12} km. This shorthand is called *standard form.*

1 Write the following in standard form:

a speed of light = 300 000 000 m/s = ______________________

b distance to nearest star = 41 000 000 000 000 km = ______________________

c furthest known galaxy = 20 000 000 000 000 000 000 km = ______________________

2 The temperature of the Sun's core is 1.4 x 10^7 °C. Write this as a normal number.

____

For each of the following, use the diagram to explain how you would do it. (You don't have to work it out unless you want to!)

The Milky Way galaxy seen from above.

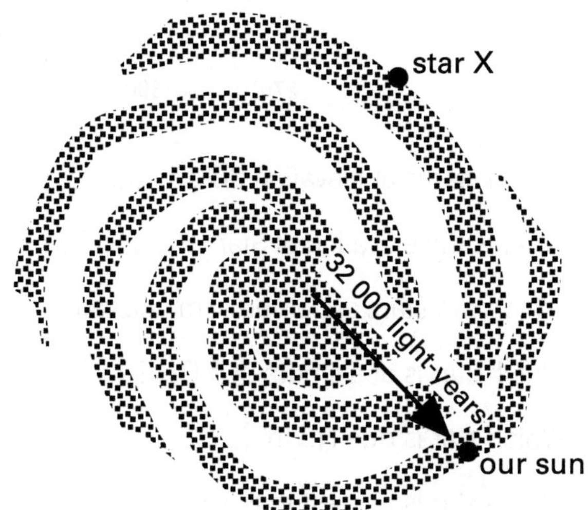

1 light-year = 9 500 000 000 000 km

3 The distance in kilometres from our Sun to the galactic centre. ____

____

4 The distance in kilometres from Star X to the galactic centre.

____

____

5 The speed an object would need to travel (in km/h) to go from the galactic centre to the Sun in 38 days.

____

____

6 The time (in hours) for a spaceship travelling at 19 000 km/h to travel from Star X to the galactic centre.

____

____

Exercise 2 Illustrating

After reading Naming stars on page 253

Imagine you discover a new constellation. Draw it in the box on the right.

What shape does it remind you of?

Use lines to show the shape.

Name your constellation and then use Bayer's system to name the four brightest stars.

Exercise 3 Using a sky grid

After doing Investigation 18 on page 255

The sky grid below shows the positions of five stars observed at approximately 8 pm.

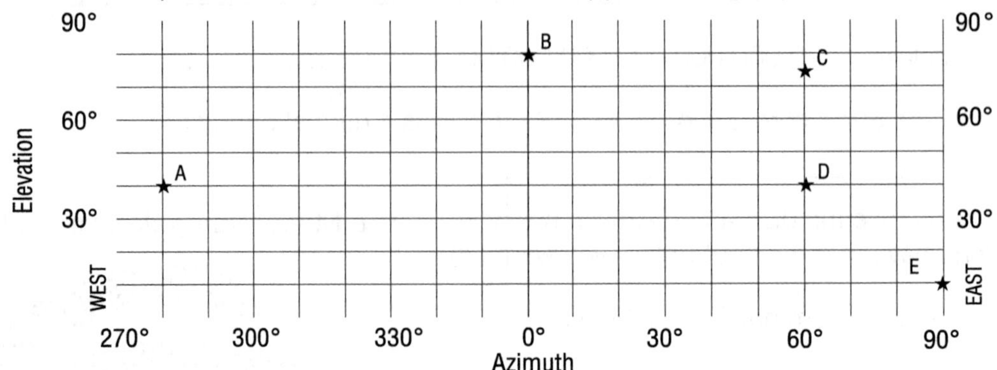

1 Which star is at elevation 80°? ______________________

2 What is the elevation of Star C? ______________________

3 Which two stars are at the same elevation? ______________________

4 Which star is at azimuth 90° East? ______________________

5 Which star is due North? ______________________

6 Which star is nearest to the zenith? ______________________

7 Which star has the smallest elevation? ______________________

8 Which two stars have the same azimuth? ______________________

9 What is the azimuth of Star A? ______________________

10 At midnight Star B has an elevation of 40° and an azimuth of 270° West. Plot its position on the grid. Why has it moved in this way? ______________________

____

Exercise 4 Recalling information

After reading pages 259–260

Use key words to complete the Acrossword on the next page. When you have finished, the letters from the shaded column will mean 'the visible part of our Sun'.

1 A state of matter that exists in the extreme heat in the interior of stars.

2 Produced in stars when hydrogen nuclei fuse.

3 A very large star, 80 000 times the size of our Sun.

4 The process by which energy in the core of the Sun moves to the surface.

5 The upper atmosphere of the Sun.

6 The scientist who reasoned that $E = mc^2$.

7 The lower atmosphere of the Sun. Its name means 'coloured ball'.

8 The gas that forms 80% of our Sun.

9 The process in which hydrogen nuclei join together to release energy.

10 Caused by columns of gas bubbling up beneath the Sun's surface.

11 Massive bursts of hot erupting gases shooting into space are solar ______________________.

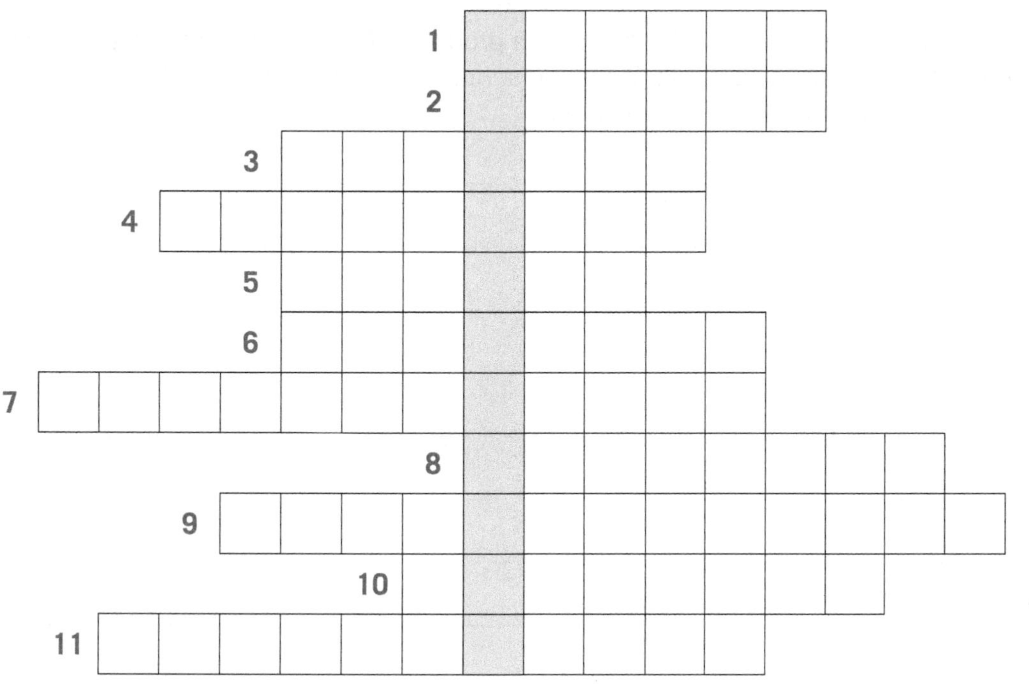

____

Exercise 5 Explaining

After reading Brightness on page 262

Kimberley is having trouble understanding the brightness of stars. Explain it to her by completing the cartoons.

This stuff about brightness of stars is crazy! If Sirius is the brightest star in the sky, why does it have a brightness of –1.4 when the dim stars have a brightness of +6?

____
____
____
____

I still don't get it. Look at the diagram on page 262. On the left, Sirius is the brightest star, but on the right it's the second dimmest. How can that be?

____
____
____

Exercise 6 Presenting a talk

After reading Binary stars on page 264

Imagine you are asked to present a three-minute talk to a group of 10-year-olds on the topic of binary stars. You decide to use two tennis balls in your talk. Write the script for your presentation, taking care to use language and ideas that 10-year-olds will understand.

____

____

____

____

____

____

____

____

Exercise 7 Matching words with meanings

After reading Life cycle of stars on pages 266–272

Match each word with its meaning.

Words	Meanings
red giant •	• a rotating neutron star that sends out pulsating radio signals
Steady State model •	• spectacular explosion that ends the life of a star
neutron star •	• the study of the origin and structure of the universe
black holes •	• last stage of a small to medium star's life when it collapses into a small dense object
protostar •	• first stage of a star's formation
cosmology •	• model that proposes that a hot, dense ball of radiation and atoms exploded about 10 billion years ago forming the universe
nebula •	• stage of a star's life when little hydrogen remains, the core shrinks, and the outer layers expand and cool
pulsar •	• massive cloud of gas and dust
white dwarf •	• formed when massive stars explode and continue to collapse, forming gravitational fields so strong that even light cannot escape
Big Bang model •	• model that proposes that the universe will stop expanding and collapse back on itself into a hot, dense ball of matter
supernova •	• model that proposes that the universe has always existed and will continue to exist
Big Crunch model •	• the collapsed core of a massive star after a supernova explosion

Roundup

Exercise 8 Problem solving using the strategy:

Select relevant information.

You and your twin sister are 15 years old. You both visit the Cosmic Travel Agency on Earth and ask about their 'Stardust' trips to neighbouring stars. You want to go to the closest binary star to the Earth, while your sister wants to go to the closest red dwarf star to the Earth.

However, because the starships travel at light speed, you need to know how old you and your sister will be when you return from your travels, and how old your 40-year-old mother will be.

The Cosmic Travel Agency gives you some travel information (listed below) to help you plan your trip.

Star	Apparent magnitude	Distance from Earth (l-y)	Characteristics
Barnard's star	9.5	6.1	red dwarf, single
Betelgeuse	0.1 to 1.0	425	red supergiant
Alpha Centauri	–0.3	4.3	yellow, binary
UV Cet	13	8.4	red dwarf, binary
Epsilon Indi	4.7	11.2	yellow dwarf, single
Procyon	0.34	11.4	yellow, binary
Regulus	1.4	91	blue-white, binary
Sirius	–1.47	8.7	white, binary
Vega	0.03	26	blue-white, single

Time for round trip (years)		Distance to star (light-years)
for crew of spaceship	for people on Earth	
2	2.1	0.24
5	6.5	1.7
7	10.7	3.8
8	14.5	5.5
9	18.7	7.4
10	24	9.8
11	30	12.7
13	45	19.4
15	80	35
40	36 000	18 000
50	420 000	210 000
60	5 000 000	2 500 000

Problem-solving process	Your solution
1 **Identify the problem**	
2 **Identify the information that you need to solve the problem (the relevant information)** Ignore all other information. Do not allow a large amount of information to confuse you.	
3 **Find the information you need**	
4 **Use only the relevant information to work out the solution to your problem**	

Chapter 11

Electrochemistry

What do you know already?

Exercise 1 Revising

Before you start this chapter, revise what you have previously learnt about chemistry.

1 Write down the symbols for the following elements:

hydrogen ______ chlorine ______ zinc ______ lead ______ silver ______

oxygen ______ aluminium ______ copper ______ iron ______ sodium ______

2 What is a copper ion? ______

3 Describe in your own words what the following equation tells you. $Zn \rightarrow Zn^{2+} + 2e^-$

______

______

______

4 Write an equation showing how chloride ions are formed from chlorine atoms.

______

5 If the rusting of iron can be described by this equation $Fe \rightarrow Fe^{3+} + 3e^-$

what is the valency of iron? ______

6 Balance the following equation: $H_2 + O_2 \rightarrow H_2O$

During reading

Exercise 2 Labelling a diagram

After reading What is an electric cell? on page 279

1 Label the electrodes and the electrolyte on this diagram of an electric cell.

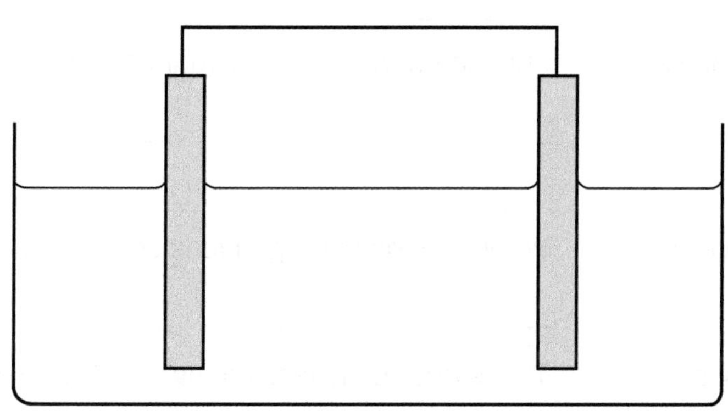

2 How is the electric current carried along the connecting wire? ____________________

3 How is the electric current carried through the liquid in the cell? ____________________

4 Write a definition of an electric cell. An electric cell is a device which ____________________

____

An example is ____________________

5 Suppose you used a solution of sodium chloride as the electrolyte. Which ions would be in the solution?

____

Exercise 3 Writing in role

After reading How an electric cell works on page 281

TASK: You are an electric cell consisting of a zinc strip and a copper strip connected by a wire and immersed in dilute sulfuric acid. Answer the following questions in an interview about how you are able to generate an electric current.

Interviewer: Today, listeners, I have with me a simple chemical cell. To start with, cell, can you tell me how much electricity you can produce?

Cell: ____________________

Interviewer: How many different parts do you have?

Cell: ____________________

____

____

Interviewer: I believe your zinc strip dissolves in the acid. It sounds horrible, but what actually happens?

Cell: ____________________

____

Interviewer: Where do the zinc ions go?

Cell: ____________________

Interviewer: And where do the electrons go?

Cell: ____________________

Interviewer: So the zinc strip becomes negatively charged. But why?

Cell: ____________________

Interviewer: I notice bubbles around your copper strip. Why is this?

Cell: ____________________

____

Interviewer: Where do the hydrogen ions come from?

Cell: ____________________

Interviewer: That's absolutely fascinating, cell. Thanks for talking with us.

Exercise 4 Summarising in a table

After reading Dry cells on page 282

1 Complete the table below that summarises the parts of a dry cell and their function. The first one has been done for you.

Part	Made of …	Function
outside case	zinc	negative electrode
carbon rod		
paste around rod		
electrolyte paste		
top of cell		

2 Complete the table showing what happens to the electrodes in a dry cell.

Electrode	Electric charge	Equation for chemical reaction	Electrons gained or lost?	Oxidation or reduction?
outside case				
carbon rod				

3 What is the difference between a cell and a battery? (Use the contrast linking word *whereas* in your answer.)

____

____

____

Exercise 5 Writing in your own words

After reading pages 282–286

1 What is the advantage of batteries that can be recharged? ____________________

2 In your own words, describe the structure of a car battery. (Do not use an illustration this time—use your word power.)

3 Name three parts of the car that operate on the discharge from the battery.

a ____________ b ____________ c ____________

4 Name the part of the car that recharges the battery.

5 In your own words, explain how the car battery can be recharged. Include these words in your explanation.

lead	electric current	sulfuric acid
lead sulfate	electrical energy	chemical energy

6 In your own words, explain how a fuel cell using hydrogen gas and air passing through electrodes can generate electricity to power an electric vehicle.

Exercise 6 Using prefixes

After reading pages 288–290

1 In this chapter, and in other chapters, you have learnt many words starting with *electro*. This prefix means *to do with* or *caused by electricity*. It comes from the Greek word *elektron*, which means amber, a natural substance that the ancient Greeks discovered became electrically charged when rubbed.

Match each electro-word in the shaded box with its meaning below.

electrochemistry	electrolysis	electromagnet	electronics
electrode	electrolyte	electron	electroplating

a A conductor that allows electric current to flow into or out of an electrolyte ____________

b The process of passing an electric current through an electrolyte to produce chemical reactions at the electrodes ____________

c A substance capable of conducting an electric current when dissolved or melted ____________

d The process of depositing a thin layer of metal on another using electrolysis ____________

e The branch of chemistry concerned with the study of electric cells and electrolysis ____________

f A tiny negatively charged subatomic particle found in the space around the nucleus of an atom ____________

g A magnet consisting of an iron or steel core wound with a coil of wire through which an electric current is passed ____________

h The science and technology concerned with devices such as diodes, resistors and transistors

____

2 By looking at the structure of the word, suggest meanings for the following:

a electrostatics ____________

____

b electrometer ____________

____

c electrotherapy ____________

____

3 Write down at least two other electro-words, with their meanings. Use a dictionary if necessary.

____

____

____

Exercise 7 Writing in role

After reading Hair removal by electrolysis on page 290

TASK: Linda is still having trouble with unwanted hairs on her face, so she has made an appointment at the local beauty salon. The treatment takes quite a while, and she discusses electrolysis with the beautician.

Use the information on page 290 to complete the discussion.

Linda: Ouch, that needle hurt! Why is it made of gold?

Beautician: ______________________________

____

Linda: Is the needle positive or negative?

Beautician: ______________________________

____

Linda: I understand why you put the needle in near the hair, but why do I have to hold this metal cylinder?

Beautician: ______________________________

Linda: What I can't understand is how the electricity gets rid of the hair.

Beautician: ______________________________

____

Linda: If it's alkaline around the needle, what happens at the other electrode that I'm holding?

Beautician: ______________________________

____

Linda: I've tried a home electrolysis kit where you hold the hair with a pair of tweezers instead of using a needle. But it didn't work very well. Why?

Beautician: ______________________________

____

Exercise 8 Recalling information

After reading Corrosion of metals on pages 293–296

1 Write the equation for the overall reaction that occurs when iron rusts.

____

2 How does painting steel prevent it from rusting? ______________________________

____

3 Use Fig 14 on page 293 to explain why the corrosion of iron is like what happens in an electric cell.

____

____

4 Use the Activity series on page 295 to decide whether each of the following statements is true or false.

a	Sodium reacts very quickly.	TRUE	FALSE
b	Silver and gold corrode more slowly than iron and steel.	TRUE	FALSE
c	Iron is not as reactive as copper.	TRUE	FALSE
d	Aluminium reacts more quickly than iron or copper.	TRUE	FALSE
e	Calcium is an unreactive metal.	TRUE	FALSE
f	Magnesium loses electrons more easily than copper does.	TRUE	FALSE
g	Silver gains electrons more easily than zinc does.	TRUE	FALSE

5 Use the Activity series to write your own correct sentence comparing the reactivity of two or more different metals. ______________________________

____

6 Complete the following sentences.

Iron is ______________ reactive than copper. So when iron and copper are in contact in salt water, the ______________ does not corrode, but the iron corrodes ______________ than normal.

Iron is ______________ reactive than magnesium. So when iron and magnesium are in contact in salt water, the magnesium ______________, but the iron ______________.

7 Look at Fig 15 on page 295. How do the blocks of zinc protect the hull of the ship from corrosion? ______________

____

____

____

8 Use Fig 16 on page 296 to explain why it is important not to use metal cans that are scratched or dented.

____

____

____

9 Aluminium is more reactive than iron, yet it does not corrode as much. How can you explain this? ______________

____

____

____

____

Roundup

Exercise 9 Problem solving

Anika wants to make a lemon battery. She has 2 lemons, 5 connecting wires, an LED and four metal strips: one iron, one magnesium, one lead and one copper.

In which order should she place the metal strips to make the best battery to light up the LED?

Which problem-solving strategy did you use? Justify your answer.

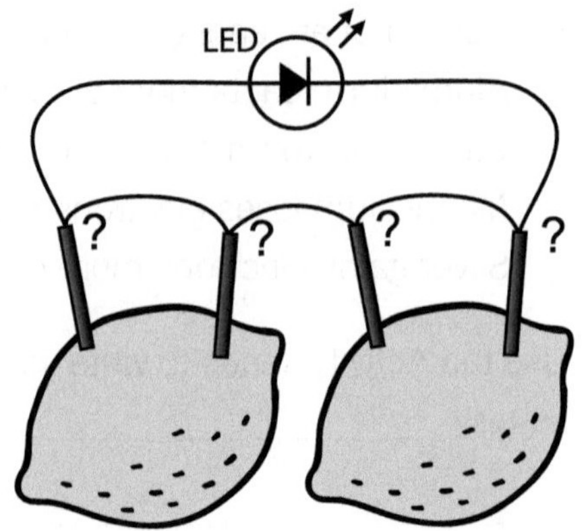

Working space

Linking words

Sequence linking words (for actions that occur one after the other)

first, second	then	before	finally	subsequently
to begin with	later	next	gradually	initially
when	from there	last	initial	after

Cause and effect linking words (for actions that cause another action to occur)

therefore	so that	an outcome of	results in
due to	as a result of	causing	consequently
since	because	caused by	this leads to
the effect of	the reason for	if … then	when … then
resulting in	making	as	accounts for
so	thereby	causes	gives rise to
creates	owing to	hence	thus

Contrast linking words (to describe differences)

on the other hand	in contrast to	different from	however
alternatively	but	although	rather than
whereas	while	unlike	yet
even though	less than	more than	whilst

Comparison linking words (to describe similarities)

both	same	in both	like
in both cases	similarly	and	as well as
in the same way	just as … so …	alike	similar in that

Generalisation linking words (for statements that are true in most cases)

generally	usually	commonly	many
normally	as a general rule	most	mostly

Notes